Fibonacci Numbers and Their Applications

Mathematics and Its Applications

Fibonacci Numbers and Their Applications

edited by

Andreas N. Philippou

University of Patras, Patras, Greece

Gerald E. Bergum

University of South Dakota,
Brookings, South Dakota, U.S.A.

and

Alwyn F. Horadam

University of New England,
Armidale, N.S.W., Australia

D. Reidel Publishing Company

A MEMBER OF THE KLUWER ACADEMIC PUBLISHERS GROUP

Dordrecht / Boston / Lancaster / Tokyo

Library of Congress Cataloging in Publication Data

International Conference on Fibonacci Numbers and their Applications
  (1st : 1984 : University of Patras)
  Fibonacci numbers and their applications.

  (Mathematics and its applications)
    1. Fibonacci numbers–Congresses. I. Bergum, Gerald E. II. Philip-
pou, Andreas N. III. Horadam, A. F. IV. Title. V. Series: Mathematics
and its applications (D. Reidel Publishing Company)
QA241.I58   1984   512'.72   86–3995

Published by D. Reidel Publishing Company
P.O. Box 17, 3300 AA Dordrecht, Holland

Sold and distributed in the U.S.A. and Canada
by Kluwer Academic Publishers,
190 Old Derby Street, Hingham, MA 02043, U.S.A.

In all other countries, sold and distributed
by Kluwer Academic Publishers Group,
P.O. Box 322, 3300 AH Dordrecht, Holland

ISBN 1-4020-0327-7
Transferred to Digital Print 2001

TABLE OF CONTENTS

EDITOR'S PREFACE

Approach your problems from the right end
and begin with the answers. Then one day,
perhaps you will find the final question.

'The Hermit Clad in Crane Feathers' in R.
van Gulik's *The Chinese Maze Murders*.

It isn't that they can't see the solution. It is
that they can't see the problem.

G.K. Chesterton. *The Scandal of Father
Brown* 'The point of a Pin'.

Growing specialization and diversification have brought a host of monographs and textbooks on increasingly specialized topics. However, the "tree" of knowledge of mathematics and related fields does not grow only by putting forth new branches. It also happens, quite often in fact, that branches which were thought to be completely disparate are suddenly seen to be related.

Further, the kind and level of sophistication of mathematics applied in various sciences has changed drastically in recent years: measure theory is used (non-trivially) in regional and theoretical economics; algebraic geometry interacts with physics; the Minkowsky lemma, coding theory and the structure of water meet one another in packing and covering theory; quantum fields, crystal defects and mathematical programming profit from homotopy theory; Lie algebras are relevant to filtering; and prediction and electrical engineering can use Stein spaces. And in addition to this there are such new emerging subdisciplines as "experimental mathematics", "CFD", "completely integrable systems", "chaos, synergetics and large-scale order", which are almost impossible to fit into the existing classification schemes. They draw upon widely different sections of mathematics. This programme, Mathematics and Its Applications, is devoted to new emerging (sub)disciplines and to such (new) interrelations as exempla gratia:

- a central concept which plays an important role in several different mathematical and/or scientific specialized areas;
- new applications of the results and ideas from one area of scientific endeavour into another;
- influences which the results, problems and concepts of one field of enquiry have and have had on the development of another.

The Mathematics and Its Applications programme tries to make available a careful selection of books which fit the philosophy outlined above. With such books, which are stimulating rather than definitive, intriguing rather than encyclopaedic, we hope to contribute something towards better communication among the practitioners in diversified fields.

A cartoon by Sidney Harris depicts a house with house number 11235813... with the caption "This must be Fibonacci's". That is how well the Fibonacci sequence of numbers $1,1,2,3,5,8,13,21,...$; $n_i = n_{i-2} + n_{i-1}$, $n_1 = n_2 = 1$, is known. The Lucas numbers $2,1,3,4,7,11,...$ are almost equally well known. There is of course a general well known closed formula for the n-th term of the Fibonacci sequence. So, all in all, at first sight, one would say that there really cannot be anything very deep, profound or interesting about them. And, frankly, that was my own initial assessment. Because the recursive rule of obtaining them is so simple there is of course the fact that the Fibonacci and Lucas numbers naturally occur all over mathematics and science. That improves things but is still not very convincing. This book, however, is. There *are* fascinating, stimulating and profound aspects to these simple natural sequences. Indeed, being recursively simple is often deceptive and often hides real depth. This book, the proceedings of the first international conference on the topic, provides a survey that goes a long way to establishing that there is a good deal of meat to the legacy of Fibonacci and Lucas and that there are many more applications of these sequences still to come.

The unreasonable effectiveness of mathematics in science ...

Eugene Wigner

Well, if you know of a better 'ole, go to it.

Bruce Bairnsfather

What is now proved was once only imagined.

William Blake

As long as algebra and geometry proceeded along separate paths, their advance was slow and their applications limited.

But when these sciences joined company they drew from each other fresh vitality and thenceforward marched on at a rapid pace towards perfection.

Joseph Louis Lagrange.

Bussum, March 1986   Michiel Hazewinkel

A REPORT ON THE FIRST INTERNATIONAL CONFERENCE ON FIBONACCI NUMBERS AND THEIR APPLICATIONS

UNIVERSITY OF PATRAS, GREECE
AUGUST 27-31, 1984

Approximately fifty mathematicians from fifteen different countries gathered in Patras, on the Peloponnesos in Greece, to exchange knowledge and thoughts on various mathematical topics all with the Fibonacci numbers as a common denominator. Professor A. N. Philippou, chairman of both the international and the local organizational committees, expressed it as follows in his remarks at the opening session: "Most will be lecturing on Number Theory, some will talk on Probability, and still others will present their results on ladder networks in Electric Line Theory and aromic hydrocarbons in Chemistry."

The academic sessions were scheduled, of course, according to the pace of the host country. A morning session from 9:00 A.M. to 1:00 P.M. and an afternoon session from 5:00 P.M. to 8:30 P.M., each session interrupted once by a coffee break. All lectures lasted for 45 minutes and all were in the nature of contributed papers—twenty-eight in total. The Conference Proceedings are published herein.

The relatively small number of participants made the conference a pleasant affair; in no time everyone knew everyone else. The social atmosphere was enhanced still more by outings and parties conducted under the friendly guidance of the Greek colleagues. Clearly, it is difficult to improve an environment that appropriately could be called the "cradle of mathematics."

At the end of the final session, Professor Philippou and his committees and staff were given well deserved praise and applause. It was suggested that similar international conferences should be held every other year, and that the University of Santa Clara, in California, U.S.A., "home of The Fibonacci Association," should be the host in 1986, followed in 1988 by an appropriate institution in Pisa, Italy, birthplace of Fibonacci.

The Conference was jointly sponsored by the Greek Ministry of Culture and Science, the Fibonacci Association, and the University of Patras.

Karel L. de Bouvère

CONTRIBUTORS

PROFESSOR PETER G. ANDERSON   (pp. 1-8)
School of Computer Science and Technology
Rochester Institute of Technology
P.O. Box 9887
Rochester, New York 14623, U.S.A.

PROFESSOR NGUYEN-HUU BONG   (pp. 9-37)
School of Mathematical Sciences
Universiti Sains Malaysia
Minden, Penang, MALAYSIA

PROFESSOR HERTA T. FREITAG   (pp. 39-41)
B40 Friendship Manor
320 Hershberger Road, N.W.
Roanoke, Virginia 24012, U.S.A.

PROFESSOR COSTAS GEORGHIOU   (pp. 229-233)
Department of General Studies
University of Patras
Patras, GREECE

PROFESSOR KATUOMI HIRANO   (pp. 43-53)
The Institute of Statistical Mathematics
4-6-7 Minami-Azabu Minato-Ku
Tokyo 106, JAPAN

PROFESSOR ALWYN HORADAM   (pp. 55-80; 81-97; 163-180)
Department of Mathematics, Statistics
and Computing Science
The University of New England
Armidale, N.S.W. 2351, AUSTRALIA

DR. DANIELA JARUŠKOVÁ   (pp. 99-104)
Na Folimance 5
Praha 2, CZECHOSLOVAKIA-ČSSR

DOZ. DR. PETER KIRSCHENHOFER   (pp. 105-120)
Institut für Algebra und Diskrete Mathematik
Technische Universität Wien
Gußhaustraße 27-29
A-1040 Wien, AUSTRIA

PROFESSOR PÉTER KISS   (pp. 121-130; 131-139)
Ho Si Minh Tanárképzö Foiskola
Matematikai Tanszék
Eger, Leányka u. 4, HUNGARY

DR. JOSEPH LAHR   (pp. 141-161)
Institut Supérieur de Technologie
LUXEMBOURG

PROFESSOR ERIK LIEUWENS   (pp. 131-139)
3817 DC Amersfoort
NETHERLANDS

BR. J. M. MAHON (pp. 55-80; 163-180)
Department of Mathematics, Statistics and Computing Science
The University of New England
Armidale, N.S.W. 2351, AUSTRALIA

PROFESSOR J. H. McCABE   (pp. 181-184)
The Mathematical Institute
University of St. Andrews
St. Andrews, SCOTLAND KY 16 9SS, U.K.

DR. S. PETHE   (pp. 185-192)
Department of Mathematics
University of Malaya
Kuala Lumpur, MALAYSIA

PROFESSOR A. PETHÖ   (pp. 193-201)
Mathematical Institute
Kossuth Lajos University
H-4010 Debrecen, Pf 12, HUNGARY

PROFESSOR ANDREAS N. PHILIPPOU   (pp. 203-227)
Department of Mathematics
University of Patras
Patras, GREECE

PROFESSOR G. N. PHILIPPOU   (pp. 229-233)
Department of Mathematics
University of Patras
Patras, GREECE

PROFESSOR G. M. PHILLIPS   (pp. 181-184)
The Mathematical Institute
University of St. Andrews
St. Andrews, SCOTLAND KY 16 9SS, U.K.

PROFESSOR BUI MINH PHONG   (pp. 131-139)
Ho Si Minh Tanárképzö Foiskola
Matematikai Tanszék
Eger, Leányka u. 4, HUNGARY

DOZ. DR. HELMUT PRODINGER   (pp. 105-120)
Institut für Algebra und Diskrete Mathematik
Technische Universität Wien
Gußhaustraße 27-29
A-1040 Wien, AUSTRIA

PROFESSOR GERHARD ROSENBERGER   (pp. 235-240)
Abteilung Mathematik
Universität Dortmund
Postfach 50 05 00
D-4600 Dortmund 50, WEST GERMANY

PROFESSOR ANDRZEJ ROTKIEWICZ   (pp. 241-255)
Instytut Matematyczny
Polskiej Akademii Nauk-PAN
Skr. Sniadeckich 8
P-00-950 Warszawa, POLAND

PROFESSOR A. G. SHANNON   (pp. 81-97)
School of Mathematical Sciences
The New South Wales Institute of Technology
Broadway, N.S.W. 2007, AUSTRALIA

PROFESSOR LAWRENCE SOMER   (pp. 257-272)
1400 20th Street, N.W. #619
Washington, D.C. 20036, U.S.A.

DOZ. FR. ROBERT F. TICHY   (pp. 105-120; 273-291)
Institut für Analysis, Technische Mathematik
und Versicherungsmathematik
Technische Universität Wien
Wien, AUSTRIA

DR. KEITH TOGNETTI   (pp. 293-304)
Department of Mathematics
The University of Wollongong
P.O. Box 1144
Wollongong, N.S.W. 2500, AUSTRALIA

PROFESSOR TONY VAN RAVENSTEIN   (pp. 293-304)
Department of Mathematics
The University of Wollongong
P.O. Box 1144
Wollongong, N.S.W. 2500, AUSTRALIA

PROFESSOR GRAHAM WINLEY   (pp. 293-304)
Department of Mathematics
The University of Wollongong
P.O. Box 1144
Wollongong, N.S.W. 2500, AUSTRALIA

FOREWORD

This book contains twenty-two papers from among the thirty-
one papers presented at the First International Conference
on Fibonacci Numbers and Their Applications. These papers
have been selected after a careful review by eminent experts
in the field, and they range from elementary number theory
and probability to electrical engineering and chemistry.
The Fibonacci numbers are their unifying bond.
    It is anticipated that the book will be useful to
research workers and graduate students interested in the
Fibonacci numbers and their applications.

December 1985   The Editors

                                     Gerald E. Bergum
                                     University of South Dakota
                                     Brookings, South Dakota, U.S.A.

                                     Andreas N. Philippou
                                     University of Patras
                                     Patras, Greece

                                     Alwyn F. Horadam
                                     University of New England
                                     Armidale, N.S.W., Australia

THE ORGANIZING COMMITTEES

LOCAL COMMITTEE

INTERNATIONAL COMMITTEE

Philippou, A., *Chairman*

Philippou, A. (Greece), *Chairman*

Metakides, G., *Cochairman*

Bergum, G. (U.S.A.), *Cochairman*

Cotsiolis, A.

Bicknell-Johnson, M. (U.S.A.)

Drossos, C.

Georghiou, C. (Greece)

Iliopoulos, D.

Horadam, A. (Australia)

Ioakimidis, N.

Kiss, P. (Hungary)

Philippou, G.

Muwafi, A. (Lebanon)

Zachariou, A.

Rotkiewicz, A. (Poland)

[†]*The asterisk indicates that the paper is included in this
book.*

*KISS, P., Teacher Training College, Eger (coauthors
    E. Lieuwens and B. M. Phong). "On Lucas Pseudoprimes
    Which Are Products of s Primes."
*LAHR, J., Institut Supérieur de Technologie, Luxembourg.
    "Fibonacci and Lucas Numbers and the Morgan-Voyce
    Polynomials in Ladder Networks and in Electric Line
    Theory."
*LIEUWENS, E. (coauthors P. Kiss and B. M. Phong). "On
    Lucas Pseudoprimes Which are Products of s Primes."
*Bro. MAHON, J. M., Catholic College of Education, Sydney
    (coauthor A. F. Horadam). "Convolutions for Pell
    Polynomials."
*Bro. MAHON, J. M., Catholic College of Education, Sydney
    (coauthor A. F. Horadam). "Infinite Series Summation
    Involving Reciprocals of Pell Polynomials."
 MARKATOS, K. M., Technological Educational Institute,
    Larissa. "The Study of the Earthquake by the Lex-
    arithmic Theory."
*McCABE, J. H., University of Saint Andrews, Saint Andrews
    (coauthor G. M. Phillips). "Fibonacci and Lucas Num-
    bers and Aitken Acceleration."
 PANARETOS, J., University of Patras, Patras (coauthors
    A. N. Philippou and E. Xekalaki). "On Some Mixtures
    of Distributions of Order k."
 PAPASTAVRIDES, S., University of Patras, Patras. "Some
    Remarks on the Discrete Distributions of Order k."
*PETHE, S., Universiti Malaya, Kuala Lumpur. "On Sequences
    Having Third-Order Recurrence Relations."
*PETHÖ, A., Kossuth Lajos Universität Debrecen, Debrecen.
    "On the Solution of the Equation $G_n = P(x)$."
*PHILIPPOU, A. N., University of Patras, Patras. "Distribu-
    tions and Fibonacci Polynomials of Order k, Longest
    Runs, and Reliability of Consecutive-k-out-of-n : F
    Systems."
 PHILIPPOU, A. N., University of Patras, Patras (coauthors
    J. Panaretos and E. Xekalaki). "On Some Mixtures of
    Distributions of Order k."
*PHILIPPOU, G. N., Higher Technical Institute, Nicosia
    (coauthor C. Georghiou). "Fibonacci-Type Polynomials
    and Pascal Triangles of Order k."
*PHILLIPS, G. M., University of Saint Andrews, Saint Andrews
    (coauthor J. H. McCabe). "Fibonacci and Lucas Numbers
    and Aitken Acceleration."

*PHONG, B. M., Teacher Training College, Eger (coauthors
    P. Kiss and E. Lieuwens). "On Lucas Pseudoprimes
    Which Are Products of s Primes."

 POPOV, B. S., Macedonian Academy of Sciences and Arts,
    Skopje. "On Reciprocal Series of Generalized Fibonacci
    Numbers."

*PRODINGER, H., Technische Universität Wien, Wien (coauthors
    P. Kirschenhofer and R. Tichy). "Fibonacci Numbers of
    Graphs III: Planted Plane Trees."

 ROBBINS, N., San Francisco State University, San Francisco.
    "On Pell Numbers of the Form px^2, Where p Is Prime."

*ROSENBERGER, G., Universität Dortmund, Dortmund. "A Note
    On Fibonacci and Related Numbers in the Theory of
    2×2 Matrices."

*ROTKIEWICZ, A., Polish Academy of Sciences, Warszawa.
    "Problems on Fibonacci Numbers and Their Generaliza-
    tions."

*SHANNON, A. G., N.S.W. Institute of Technology, Sydney
    (coauthor A. F. Horadam). "Cyclotomy-Generated Poly-
    nomials of Fibonacci Type."

*SOMER, L., George Washington University, Washington.
    "Linear Recurrences Having Almost All Primes as Maxi-
    mal Divisors."

*TICHY, R., Technische Universität Wien, Wien. "On the
    Asymptotic Distribution of Linear Recurrence Sequences."

*TICHY, R., Technische Universität Wien, Wien (coauthors
    P. Kirschenhofer and H. Prodinger). "Fibonacci Numbers
    of Graphs III: Planted Plane Trees."

*TOGNETTI, K., University of Wollongong, Wollongong (co-
    authors T. Van Ravenstein and G. Winley). "Golden
    Hops Around a Circle."

 TSANGARIS, P., University of Athens, Athens. "Prime Num-
    bers in General and of Particular Form."

*VAN RAVENSTEIN, T., University of Wollongong, Wollongong
    (coauthors K. Tognetti and G. Winley). "Golden Hops
    Around a Circle."

*WINLEY, G., University of Wollongong, Wollongong (coauthors
    K. Tognetti and T. Van Ravenstein). "Golden Hops
    Around the Circle."

 XEKALAKI, E., University of Crete, Heraklion (coauthors
    J. Panaretos and A. N. Philippou). "On Some Mixtures
    of Distributions of Order k."

 ZACHARIOU, A., University of Athens, Athens. "Squaring
    the Circle? Let Us Drink to It."

INTRODUCTION

The numbers

$$1, 1, 2, 3, 5, 8, 13, 21, 34, 55, 89, \ldots,$$

known as Fibonacci numbers have been named by the nineteenth-century French mathematician Edouard Lucas after Leonard Fibonacci of Pisa, one of the best mathematicians of the Middle Ages, who referred to them in his book *Liber Abaci* (1202) in connection with his rabbit problem.

The astronomer Johann Kepler rediscovered the Fibonacci numbers, independently, and since then several renowned mathematicians have dealt with them. We only mention a few: J. Binet, B. Lamé, and E. Catalan. Edouard Lucas studied Fibonacci numbers extensively, and the simple generalization

$$2, 1, 3, 4, 7, 11, 18, 29, 47, 76, 123, \ldots,$$

bears his name.

During the twentieth century, interest in Fibonacci numbers and their applications rose rapidly. In 1961 the Soviet mathematician N. Vorobyov published *Fibonacci Numbers*, and Verner E. Hoggatt, Jr., followed in 1969 with his *Fibonacci and Lucas Numbers*. Meanwhile, in 1963, Hoggatt and his associates founded The Fibonacci Association and began publishing *The Fibonacci Quarterly*. They also organized a Fibonacci Conference in California, U.S.A., each year for almost sixteen years until 1979. In 1984, the First International Conference on Fibonacci Numbers and Their Applications was held in Patras, Greece, and it was anticipated then that it would set the beginning of international conferences on the subject to be held every two or three years in different countries. As a matter of fact, the Second International Conference on Fibonacci Numbers and Their Applications is planned to take place in San Jose, California, U.S.A., August 13-16, 1986.

It is impossible to overemphasize the importance and
relevance of the Fibonacci numbers to the mathematical
sciences and other areas of study. The Fibonacci numbers
appear in almost every branch of mathematics, like number
theory, differential equations, probability, statistics,
numerical analysis, and linear algebra. They also occur in
biology, chemistry, and electrical engineering.

It is believed that the contents of this book will
prove useful to everyone interested in this important branch
of mathematics and that they may lead to additional results
on Fibonacci numbers both in mathematics and in their applications in science and engineering.

The Editors

Peter G. Anderson

FIBONACCENE

INTRODUCTION

Yet another physical phenomenon is exposed in which the
Fibonacci numbers and the golden mean play a prominent role;
in this case, it is a family of aromatic hydrocarbons. The
importance of the present results for Chemistry is yet to be
determined, but insights are gained into the theory of
matching, and a natural discovery of a recurrence formula
for the Fibonacci numbers is presented.

THE CHEMISTRY OF CHICKEN-WIRE:
AROMATIC HYDROCARBONS

A glance at the figures that accompany this article will
show the reader the category of organic chemical compounds
we are considering, namely, polycyclic aromatic hydrocarbons.
The molecules consist of fused benzene rings that form a
planar graph (or the reader may prefer to think of them as
finite subgraphs of an infinite chicken-wire fence).

MATCHINGS IN GRAPHS: KEKULÉ STRUCTURES
FOR MOLECULES

The rules of chemical valences imply that every carbon atom
in our molecules must be associated with exactly one double
bond. In the terminology of graph theory, we require a
"matching," which is a subset of the edges of the given
graph such that every vertex of the graph is an end-point
of exactly one of the chosen edges. Organic chemists call
such a matching in aromatic hydrocarbons a "Kekulé structure."
    Figure 2 shows the two possible Kekulé structures for
benzene, and three for naphthalene.

1

A. N. Philippou et al. (eds.), Fibonacci Numbers and Their Applications, 1–8.
© 1986 by D. Reidel Publishing Company.

One useful purpose this Kekulé structure enumeration serves is to classify proposed molecular structures as possible or impossible. For example, the three forms shown in Figure 3 are all impossible compounds. Each admits no Kekulé structure. Figure 3a has an odd number of vertices, so no matching is possible. Figure 3b illustrates a deeper principle. The orientation of our drawings suggests that we may label carbons as either Y or L (for lambda) types (see Figure 4). Any matching must pair every Y-carbon with an L-carbon, but Figure 3b has thirteen Y-carbons and only eleven L-carbons.* The impossibility of Figure 3c is a consequence of the heavier theory of matching in graphs (e.g., the principle of Hungarian trees [Christophedes] or unaugmentable matchings [Harary]), which we will not pursue further here.

Figure 5 shows a generic example of a Kekulé structure of n-benzene (a straight line of n fused benzine rings; here $n = 9$). Note that there will always be exactly one of the vertical bonds chosen to be double in any matching, and, consequently, n-benzene has $n + 1$ Kekulé structures.

FIBONACCENE

Figure 6 shows the first few members of the family or aromatic hydrocarbons that we have chosen to call "fibonaccenes." The reader may verify that the first three members of the sequence (aka benzene, naphthalene, phenanthrene, and chrysene) admit 2, 3, 5, and 8 Kekulé structures, respectively—a short but recognizable sequence. To count the Kekulé structures of n-fibonaccene (n fused benzene rings), proceed as follows (we use 9-fibonaccene, Figure 7). If the bond labeled b is double (in the matching) then the four other bonds that belong exclusively to the first ring are determined, but no restrictions are placed on the choice of double bonds of the remaining $n - 1$ rings. If bond b is single (not in the matching), then choices are forced upon us for all ten bonds in the first two rings, and no

---

This is similar to the well-known impossible puzzle: remove two diagonally opposite squares from a checker board and attempt to cover the remaining 62 squares with 31 one-by-two dominoes.

restrictions are placed on the bonds of the remaining $n - 2$
rings. Denoting by $K(n)$ the number of Kekulé structures on
n-fibonaccene, we conclude:

$$K(1) = 2, \qquad K(2) = 3,$$
$$K(n) = K(n - 1) + K(n - 2), \text{ for } n > 2.$$

So, $K(n) = F(n + 2)$, the $(n + 2)^{nd}$ Fibonacci number.

Assuming that the bonds in an aromatic molecule are the
average of all the different possible Kekule structures
(this is naive chemistry, but it approximates an explanation
to some phenomena), we find that four of the bonds in the
rings at the ends of n-benzene are almost double bonds (the
bond labeled b in Figure 5 has an average bond strength
equal to 1.9). This is in fact where n-benzene shows its
instability; the heavy concentration of electrons causes it
to react at the ends. Following the same reasoning, the
bond labeled b in Figure 7 has an average strength of

$$\frac{K(n) + K(n + 1)}{K(n)}$$

which approximates the golden mean of 1.618..., so fibonaccene is not too active at its ends. Next, we investigate
bonds near the middle of n-fibonaccene.

DOUBLE BONDS IN THE MIDDLE:
A RECURRENCE FORMULA

Denoting $n = p + q + 1$, we shall consider bonds in the
$(p + 1)^{st}$ ring in n-fibonaccene. Let c denote a bond in
the $(p + 1)^{st}$ ring in the position shown in Figure 7. As
before, c is either a double bond or a single bond. If c is
a double bond, then both of its immediate neighbors must be
single bonds, and no restriction is placed on either the
p-fibonaccene or the q-fibonaccene molecules that precede
or follow our chosen ring. [Note that in this case, the
bond directly opposite c in the $(p + 1)^{st}$ ring must be a
single bond.] If c is a single bond, then the single/double
choices are forced for fourteen bonds in rings p, $p + 1$, and
$p + 2$, but no restrictions are placed on the bonds in the
remaining $(p - 1)$- or $(q - 1)$-fibonaccene molecules. The
above reasoning culminates in the recurrence relation:

$$K(p + q + 1) = K(p)K(q) + K(p - 1)K(q - 1).$$

[For the Fibonacci sequence, this recurrence, of course, becomes

$$F(p + q + 1) = F(p + 1)F(q - 1) + F(p)F(q),$$

which can be proved easily from the basic definition. Incidentally, student programmers who find the classical Fibonacci recursive function example too inefficient can base a very efficient recursive program on the above formula.]

The strength of the chemical bond at c is then found to be

$$\frac{2K(p)K(q) + K(p - 1)K(q - 1)}{K(p + q + 1)}.$$

APPLICATION OF FORMAL LANGUAGE THEORY

Consider the bonds in a general fibonaccene molecule: in each hexagon (except at the two ends) there is a unique bond whose two carbon atoms belong exclusively to that hexagon (the correct corresponding bond in the end hexagons can be deduced). Choose a Kekulé structure for a fibonaccene molecule, and label the hexagons in that molecule with a C if the distinguished bond is double and with a D if the bond is single (this has been done in Figure 9). Because of the rules for Kekulé structures and matchings, we see that:

(1) a sequence of n C's and D's describes a Kekulé structure for n-fibonaccene completely, and

(2) all sequences are possible *except* sequences that have two D's in a row.

The second observation above can be encoded in the notation of formal language theory by the following "regular expression,"

$$R = (C + DC)^*(D + e),$$

in which R is defined to be the "language of legal code words," "+" indicates a choice, "*" indicates a repetition zero or more times, and "e" denotes the null string. The first few members of this "language" are:

$$e$$
$$C$$
$$D$$
$$CC$$
$$CD$$
$$DC$$
$$CCC$$
$$DCC$$
$$CDC$$
$$CCD$$
$$DCD$$
$$CCCC$$
$$DCCC$$
$$CDCC$$
$$CCDC$$
$$CCCD$$
$$CDCD$$
$$DCCD$$
$$DCDC$$

Note that we have

1 string of length 0,
2 strings of length 1,
3 strings of length 2,
5 strings of length 3,
8 strings of length 4.

We can use lists such as those above to evaluate the average bond strengths in fibonaccene molecules. Give C the value 2 and D the value 1. When we examine the eight code words for 4-fibonaccene (chrysene), we get the following average values of the valences of the distinguished bonds,

13/8, 14/8, 14/8, 13/8,

and from this we can derive the average valences for all the bonds of the molecule.

The Fibonacci rule for the number of strings is easily seen in the following formation rules:

(1) If W is any valid code word of length $n - 1$, then WC is a valid code word of length n;

(2) If X is any valid code word of length $n - 2$, then XCD is a valid code word of length n;

(3) Every valid code word of length n is formed
according to rule (1) or rule (2) above.

This code word formation is described graphically by the
finite-state machine or transition network in Figure 8.

SOME CHEMISTRY COMMENTS

Students beginning their study of Chemistry (or Graph
Theory) are often mystified that chemical isomers have different properties (respectively, that graphs with the same
vertex and edge count are not isomorphic). The constrast of
n-benzene with n-fibonaccene should provide an interesting
example.
     The isomeric compounds 9-benzene and 9-fibonaccene
have, respectively, 10 and 89 Kekulé structures. (This
could be restated in terms of Johnson structures or "aromatic sextets" [Klar].) What properties do they share?
What properties differ?
     This is not the first time the Fibonacci numbers have
been identified in these aromatic hydrocarbons; see [Gordon].
For related discussions on the applications of graph theory
to chemistry, see [Trinajstić].

FIGURES

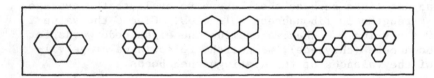

Figure 1. Some chicken-wire compounds

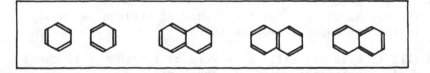

Figure 2. The Kekulé structures for benzene and naphthalene

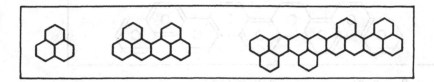

Figure 3.   Three impossible compounds

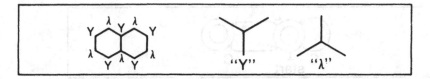

Figure 4.   Y-carbons and L-carbons

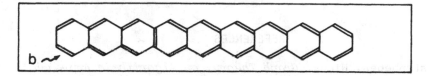

Figure 5.   A Kekulé structure on 9-benzene

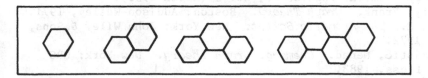

Figure 6.   A few members of the fibonaccene family

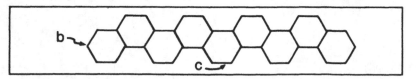

Figure 7.   9-fibonaccene

8 P. G. ANDERSONP. G. ANDERSON

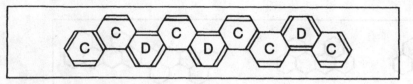

Figure 8.   9-Fibonaccene with Kekulé structure corresponding
to the code word "CCDCDCCDC."

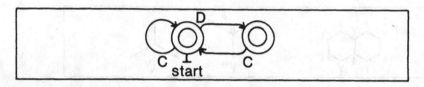

Figure 9.   A finite state machine which generates valid
Kekulé code words for n-Fibonaccene.

REFERENCES

Christophedes, Nicos. *Graph Theory, an Algorithmic Approach.*
New York: Academic Press, 1975.
Gordon, M., and Davison, W. H. T.   'Theory of Resonance
Topology of Fully Aromatic Hydrocarbons.' *The Journal
of Chemical Physics* 20, no. 3 (March 1952).
Harary, Frank. *Graph Theory.* Boston: Addison-Wesley, 1971.
Klar, E. *The Aromatic Sextet.* New York: John Wiley & Sons,
1972.
Trinajstić, Nenad. *Chemical Graph Theory.* New York: CRC
Press, 1983.Christophedes, Nicos. *Graph Theory, an Algorithmic Approach.*
New York: Academic Press, 1975.
Gordon, M., and Davison, W. H. T.   'Theory of Resonance
Topology of Fully Aromatic Hydrocarbons.' *The Journal
of Chemical Physics* 20, no. 3 (March 1952).
Harary, Frank. *Graph Theory.* Boston: Addison-Wesley, 1971.
Klar, E. *The Aromatic Sextet.* New York: John Wiley & Sons,
1972.
Trinajstić, Nenad. *Chemical Graph Theory.* New York: CRC
Press, 1983.

Nguyen-Huu Bong

ON A CLASS OF NUMBERS RELATED TO BOTH THE FIBONACCI AND PELL NUMBERS

1. INTRODUCTION

The numbers we are going to consider appeared in a paper by Entringer and Slater [4] who introduced them while investigating the problem of information dissemination through telegraphs. These numbers satisfy the following recurrence relation

$$R(k, n) = kR(k, n - 1) + R(k, n - 2), \left.\begin{array}{c} \\ \\ \end{array}\right\}$$

$$R(k, 0) = 0, \ R(k, 1) = 1, \qquad (1.1)$$

for integers $k \geqslant 1$ and $n \geqslant 2$. Observe that

(i) for $k = 1$, we have

$$R(1, n) = F_n,$$

the Fibonacci numbers, and (1.1) becomes

$$F_n = F_{n-1} + F_{n-2}, \left.\begin{array}{c} \\ \\ \end{array}\right\}$$

$$F_1 = 1 = F_2, \ F_0 = 0; \qquad (1.2)$$

(ii) for $k = 2$, we have

$$R(2, n) = P_n,$$

the Pell numbers that satisfy

$$P_n = 2P_{n-1} + P_{n-2}, \left.\begin{array}{c} \\ \\ \end{array}\right\}$$

$$P_0 = 0, \ P_1 = 1. \qquad (1.3)$$

In what follows, we shall investigate some of the properties of the numbers $R(k, n)$ as recursively defined by (1.1). As we shall presently see, iteration and

9

A. N. Philippou et al. (eds.), Fibonacci Numbers and Their Applications, 9–37.
© 1986 by D. Reidel Publishing Company.

manipulation of the relation (1.1) will lead to a multitude of results, of which known results concerning the Fibonacci and Pell numbers are special cases.

2. SOME PRELIMINARY RESULTS

From the recurrence relation (1.1), we can write, in succession,

$$
\begin{aligned}
kR(k,\ 1) &= R(k,\ 2) &&- R(k,\ 0), \\
kR(k,\ 2) &= R(k,\ 3) &&- R(k,\ 1), \\
kR(k,\ 3) &= R(k,\ 4) &&- R(k,\ 2), \\
kR(k,\ 4) &= R(k,\ 5) &&- R(k,\ 3) \\
&\ \ \vdots &&\ \ \vdots \\
kR(k,\ n-1) &= R(k,\ n) &&- R(k,\ n-2), \\
kR(k,\ n) &= R(k,\ n+1) &&- R(k,\ n-1).
\end{aligned}
$$

On adding, we obtain

$$
k \sum_{p=0}^{n} R(k,\ p) = R(k,\ n+1) + R(k,\ n) - 1, \qquad (2.1)
$$

for $k \geqslant 1$ and $n \geqslant 0$.

In a similar way, by writing in succession

$$
kR(k,\ 2p-1) = R(k,\ 2p) - R(k,\ 2p-2)
$$

and then

$$
kR(k,\ 2p) = R(k,\ 2p+1) - R(k,\ 2p-1)
$$

for $p = 1(1)n$, i.e., for $p = 1, 2, 3, \ldots, n$, and adding, we obtain

$$
k \sum_{p=1}^{n} R(k,\ 2p-1) = R(k,\ 2n) \qquad (2.2)
$$

and

$$
1 + \sum_{p=1} R(k,\ 2p) = R(k,\ 2n+1), \qquad (2.3)
$$

respectively.

Relations (2.2) and (2.3) combined give

$$
k \sum_{p=1}^{2n} (-1)^{p-1} R(k,\ p) = 1 + R(k,\ 2n) - R(k,\ 2n+1).
$$

$$
(2.4)
$$

In addition to (2.2) and (2.3), there are also relations which express both $R(k, 2n)$ and $R(k, 2n-1)$ as linear combinations of the $R(k, p)$, $p = 0(1)n$. In fact, repeated applications of (1.1) give:

$$R(k, 2n) = kR(k, 2n-1) + R(k, 2n-2)$$

$$= \binom{1}{0}R(k, 2n-2) + k\binom{1}{1}R(k, 2n-1)$$

$$= \binom{1}{0}(kR(k, 2n-3) + R(k, 2n-4)) +$$

$$+ k\binom{1}{1}(kR(k, 2n-2) + R(k, 2n-3))$$

$$= \binom{1}{0}R(k, 2n-4) + k\left(\binom{1}{0} + \binom{1}{1}\right)R(k, 2n-3) +$$

$$+ k^2\binom{1}{1}R(k, 2n-2)$$

$$= \binom{2}{0}R(k, 2n-4) + k\binom{2}{1}R(k, 2n-3) +$$

$$+ k^2\binom{2}{2}R(k, 2n-2)$$

$$= \binom{2}{0}(kR(k, 2n-5) + R(k, 2n-6)) +$$

$$+ k\binom{2}{1}(kR(k, 2n-4) + R(k, 2n-5)) +$$

$$+ k^2\binom{2}{2}(kR(k, 2n-3) + R(k, 2n-4))$$

$$= \binom{2}{0}R(k, 2n-6) + k\left(\binom{2}{0} + \binom{2}{1}\right)R(k, 2n-5) +$$

$$+ k^2\left(\binom{2}{1} + \binom{2}{2}\right)R(k, 2n-4) + k^3\binom{2}{2}R(k, 2n-3)$$

$$= \binom{3}{0}R(k, 2n-6) + k\binom{3}{1}R(k, 2n-5) +$$

$$+ k^2\binom{3}{2}R(k, 2n-4) + k^3\binom{3}{3}R(k, 2n-3).$$

Continuing the process, we obtain

$$R(k, 2n) = \sum_{p=0}^{n} k^p\binom{n}{p}R(k, p) \qquad (2.5)$$

with

$$F_{2n} = \sum_{p=0}^{n} \binom{n}{p} F_p \quad \text{and} \quad P_{2n} = \sum_{r=0}^{n} 2^r \binom{n}{r} P_r$$

as corresponding results for the Fibonacci and Pell numbers, respectively.

To obtain the corresponding relation for $R(k, 2n+1)$, we shall make use of both (1.1) and (2.5). Indeed, we have:

$$\begin{aligned}
R(k, 2n+1) &= kR(1, 2n) + R(k, 2n-1) \\
&= kR(k, 2n) + kR(k, 2n-1) + R(k, 2n-3) \\
&= kR(k, 2n) + kR(k, 2n-2) + kR(k, 2n-4) + \\
&\quad + R(k, 2n-5) \\
&= kR(k, 2n) + kR(k, 2n-2) + kR(k, 2n-4) + \\
&\quad + kR(k, 2n-6) + R(k, 2n-7).
\end{aligned}$$

Continuing the process, we obtain

$$R(k, 2n+1) = k \sum_{p=0}^{n} R(k, 2n - 2p) + R(k, 1),$$

noting that $R(k, 0) = 0$ and $R(k, 1) = 1$.

Now, by (2.5),

$$\begin{aligned}
\sum_{p=0}^{n} R(k, 2n - 2p) &= \sum_{p=0}^{n} R(k, 2(n-p)) \\
&= \sum_{p=0}^{n} \sum_{s=0}^{n-p} k^s \binom{n-p}{s} R(k, s) \\
&= \sum_{s=0}^{n} k^s \sum_{p=0}^{n} \binom{n-p}{s} R(k, s) \\
&= \sum_{s=0}^{n} k^s \binom{n+1}{s+1} R(k, s),
\end{aligned}$$

since it can be shown that

$$\sum_{p=0}^{n} \binom{n-p}{s} = \binom{n+1}{s+1}.$$

Hence,

$$R(k, 2n+1) = 1 + \sum_{p=0}^{n} k^{p+1} \binom{n+1}{p+1} R(k, p), \qquad (2.6)$$

with

$$F_{2n+1} = 1 + \sum_{p=0}^{n} \binom{n+1}{p+1} F_p,$$

and

$$P_{2n+1} = 1 + \sum_{r=0}^{n} 2^{r+1} \binom{n+1}{r+1} P_r$$

as corresponding results for the Fibonacci and Pell numbers, respectively.

Let us go back once again to the defining recurrence relation (1.1). If we multiply both sides of (1.1) by $R(k, n-1)$, so that

$$kR^2(k, n-1) = R(k, n)R(k, n-1) -$$

$$- R(k, n-1)R(k, n-2),$$

then, from

$$kR^2(k, 1) \quad = R(k, 2)R(k, 1)$$
$$kR^2(k, 2) \quad = R(k, 3)R(k, 2) \quad - R(k, 2)R(k, 1)$$
$$kR^2(k, 3) \quad = R(k, 4)R(k, 3) \quad - R(k, 3)R(k, 2)$$
$$\vdots \qquad\qquad \vdots \qquad\qquad\qquad \vdots$$
$$kR^2(k, n-1) = R(k, n)R(k, n-1) - R(k, n-1)R(k, n-2)$$
$$kR^2(k, n) \quad = R(k, n+1)R(k, n) - R(k, n)R(k, n-1),$$

we obtain, on adding,

$$k \sum_{p=1}^{n} R^2(k, p) = R(k, n)R(k, n+1), \qquad (2.7)$$

with

$$\sum_{p=1}^{n} F_p^2 = F_n F_{n+1} \quad \text{and} \quad 2 \sum_{r=1}^{n} P_r^2 = P_n P_{n+1}$$

as corresponding results for the Fibonacci and Pell numbers, respectively.

We note that the results obtained above, which correspond to (2.1)-(2.5) and to (2.7) for the case in which $k = 1$, i.e., for the Fibonacci numbers, are well known and are due to Lucas (see [3]).

3. AN ANALYTIC FORMULA FOR $R(k, n)$

Let $R_k(t)$ be the ordinary generating function for the sequence $(R(k, n))$, so that, for fixed k,

$$R_k(t) = \sum_{n \geq 0} R(k, n) t^n. \tag{3.1}$$

Then, from the recurrence relation (1.1), we obtain

$$\sum_{n \geq 2} R(k, n) t^n = \sum_{n \geq 2} k R(k, n-1) t^n + \sum_{n \geq 2} R(k, n-2) t^n,$$

that is,

$$R_k(t) - t = t k R_k(t) + t^2 R_k(t),$$

whence

$$R_k(t) = t/(1 - kt - t^2) \tag{3.2}$$

where, if $k = 1$, $t/(1 - t - t^2)$ is readily recognized as the ordinary generating function for the Fibonacci sequence (F_n).

Making use of the standard partial fraction technique, we can split and expand the right side of (3.2) to obtain

$$R(k, n) = \frac{1}{\sqrt{k^2 + 4}} \left[\left(\frac{k + \sqrt{k^2 + 4}}{2} \right)^n - \left(\frac{k - \sqrt{k^2 + 4}}{2} \right)^n \right]$$

$$\tag{3.3}$$

which, written in the form

$$R(k, n) = \frac{1}{2^{n-1}} \left[\binom{n}{1} k^{n-1} + \binom{n}{3} k^{n-3} (k^2 + 4) + \right.$$

$$\left. + \binom{n}{5} k^{n-5} (k^2 + 4)^2 + \cdots \right],$$

shows that

$$2^{n-1} \left| \left[\binom{n}{1} k^{n-1} + \binom{n}{3} k^{n-3} (k + 4) + \right. \right.$$

$$\left. \left. + \binom{n}{5} k^{n-5} (k^2 + 4)^2 + \cdots \right] \right.$$

for all k, $n \geq 1$. That

$$2^{n-1} \left| \left(\sum_{r \geq 0} \binom{n}{2r + 1} 5^r \right) \right.$$

is well known (see [7]).

When $k = 2$, (3.3) becomes

$$R(2, n) = \frac{1}{2^{n-1}}\left[\binom{n}{1}2^{n-1} + \binom{n}{3}2^{n-3} \cdot 2^3 + \right.$$

$$\left. + \binom{n}{5}2^{n-5} \cdot 2^6 + \cdots\right]$$

$$= \binom{n}{1} + \binom{n}{3}2 + \binom{n}{5}2^2 + \cdots,$$

so that, for $n \geqslant 1$, the Pell numbers P_n can be written in the form

$$P_n = \sum_{r=0}^{[(n-1)/2]}\binom{n}{2r+1}2^r$$

where $[x]$ denotes, as usual, the integral part of x.

From the preceding discussion, it is clear that we should be able to write $R(k, n)$ as a polynomial of degree $(n-1)$ in k over the integral domain $\mathbb{Z}$.

In fact, using the recurrence relation (1.1) once again, we obtain, as

coefficient of t, 1

coefficient of t^2, k

coefficient of t^3, $1 + k^2$

coefficient of t^4, $2k + k^3$

coefficient of t^5, $1 + 3k^2 + k^4$

coefficient of t^6, $3k + 4k^3 + k^5$

coefficient of t^7, $1 + 6k^2 + 5k^4 + k^6$

coefficient of t^8, $4k + 10k^3 + 6k^5 + k^7$

coefficient of t^9, $1 + 10k^2 + 15k^4 + 7k^6 + k^8$

coefficient of t^{10}, $5k + 20k^3 + 21k^5 + 8k^7 + k^9$

and so on, which can be symbolically represented by the following array of numbers:

	$k^0 \equiv 1$	k	k^2	k^3	k^4	k^5	k^6	k^7	k^8	k^9	k^{10}	
t	1	0	0	0	0	0	0	0	0	0	0	...
t^2	0	1	0	0	0	0	0	0	0	0	0	...
t^3	1	0	1	0	0	0	0	0	0	0	0	...
t^4	0	2	0	1	0	0	0	0	0	0	0	...
t^5	1	0	3	0	1	0	0	0	0	0	0	...
t^6	0	3	0	4	0	1	0	0	0	0	0	...
t^7	1	0	6	0	5	0	1	0	0	0	0	...
t^8	0	4	0	10	0	6	0	1	0	0	0	...
t^9	1	0	10	0	15	0	7	0	1	0	0	...
t^{10}	0	5	0	20	0	21	0	8	0	1	0	...
t^{11}	1	0	15	0	35	0	28	0	9	0	1	...

We note that if we read the numbers in this array

(i) transversely from left to right, then the nonzero
transversals give us precisely Pascal's triangle;

(ii) along nonzero diagonals below and including the main
diagonal, then we have the following array of numbers,
known as *Tartaglia's rectangle*:

```
1  1   1   1    1    1     1      1      1      1      1   ...
1  2   3   4    5    6     7      8      9     10     11   ...
1  3   6  10   15   21    28     36     45     55     66   ...
1  4  10  20   35   56    84    120    165    220    286   ...
1  5  15  35   70  126   210    330    495    715   1001   ...
1  6  21  56  126  252   462    792   1287   2002   3003   ...
1  7  28  84  210  462   924   1716   3003   5005   8008   ...
1  8  36 120  330  792  1716   3432   6435  11440  19448   ...
1  9  45 165  495 1287  3003   6435  12870  24310  43758   ...
1 10  55 220  715 2002  5005  11440  24310  48620  92378   ...
```

where with both rows and columns numbered from zero
(starting from the top and the leftmost, respectively),
a number $A(m, n)$ in row m and column n can be obtained
from the recurrence relation

$$A(m, n) = A(m, n - 1) + A(m - 1, n)$$
$$A(0, n) = 1 = A(m, 0).$$
(3.4)

By using generating functions, we can easily show (see [1]) that

$$A(m, n) = \binom{m + n}{n} = \binom{m + n}{m}$$
(3.5)

for all $m, n \geqslant 0$.

Now, if we compare Tartaglia's rectangle and our array of numbers, then the following results are immediate, according to (3.5):

(i)   the numerical coefficient of k^{n-1} in the polynomial coefficient of t^n on the right side of (3.1) is

$$1 = A(0, n - 1), \quad (n \geqslant 1);$$

(ii)   the numerical coefficient of k^{n-3} in the polynomial coefficient of t^n on the right side of (3.1) is

$$\sum_{1}^{n-2} 1 = n - 2 = \binom{n - 2}{1} = A(1, n - 3), \quad (n \geqslant 3);$$

(iii) the numerical coefficient of k^{n-5} in the polynomial coefficient of t^n on the right side of (3.1) is

$$\sum_{n=3}^{n-2} (n - 2) = \frac{(n - 4)(n - 3)}{2} = \binom{n - 3}{2}$$
$$= A(2, \ n - 5), \quad (n \geqslant 5);$$

(iv)   the numerical coefficient of k^{n-7} in the polynomial coefficient of t^n on the right side of (3.1) is

$$\sum_{n=5}^{n-2} \binom{n - 3}{2} = \binom{n - 4}{3} = A(3, n - 7), \quad (n \geqslant 7);$$

(v)   in general, the numerical coefficient of $k^{n-(2p-1)}$ in the polynomial coefficient of t^n on the right side of (3.1) is given by

$$A(p - 1, n - (2p - 1)) = \binom{n - p}{p - 1}, \quad (p \geqslant 1),$$

all other numerical coefficients of k^8 in the polynomial coefficient of t^n on the right side of (3.1) being
zero.

It follows that

$$R(k, n) = \sum_{p=1}^{[(n+1)/2]} \binom{n-p}{p-1} k^{n-2p+1},$$

or equivalently,

$$R(k, n) = \sum_{r=0}^{[(n-1)/2]} \binom{n-r-1}{r} k^{n-r-1} \tag{3.6}$$

with

(i) $$P_n = \sum_{r=0}^{[(n-1)/2]} \binom{n-r-1}{r} 2^{n-r-1}$$

as the corresponding formula for the Pell numbers P_n,
and

(ii) $$F_n = \sum_{r=0}^{[(n-1)/2]} \binom{n-r-1}{r}$$

as the corresponding (well-known) formula for the
Fibonacci numbers F_n.

Now, let

$$a = \frac{k + \sqrt{k^2 + 4}}{2} \quad (a > 1).$$

Then,

$$a^{-1} = \frac{2}{k + \sqrt{k^2 + 4}} = -\frac{k - \sqrt{k^2 + 4}}{2},$$

so that we can write (3.3) in the form

$$R(k, n) = \frac{1}{\sqrt{k^2 + 4}} (a^n + (-1)^{n+1} a^{-n}).$$

But then,

(a) $(a^{1-n}\sqrt{k^2 + 4})R(k, n) = a + \dfrac{(-1)^{n+1}}{a^{2n-1}}$;

suppose

$$a + \frac{(-1)^{n+1}}{a^{2n-1}} \geqslant \sqrt{k^2 + 4},$$

that is,

$$\frac{(-1)^{n+1}}{a^{2n-1}} \geqslant \sqrt{k^2 + 4} - a = a^{-1};$$

then

$$\frac{(-1)^{n+1}}{a^{2n-2}} \geqslant 1,$$

which is not possible for $n \geqslant 2$; hence, for $n \geqslant 2$ and $k \geqslant 1$,

$$(a^{1-n}\sqrt{k^2 + 4})R(k, n) < \sqrt{k^2 + 4},$$

that is,

$$R(k, n) < a^{n-1} = \left(\frac{k + \sqrt{k^2 + 4}}{2}\right)^{n-1};$$

(b) $(a^{2-n}\sqrt{k^2 + 4})R(k, n) = a^2 + \dfrac{(-1)^{n+1}}{a^{2n-2}};$

suppose

$$a^2 + \frac{(-1)^{n+1}}{a^{2n-2}} \leqslant 2a - 1;$$

then

$$\frac{(-1)^{n+1}}{a^{2n-2}} \leqslant -(a^2 - 2a + 1),$$

or

$$\frac{(-1)^n}{a^{2n-2}} \geqslant (a - 1)^2,$$

which is not possible, as can be readily seen for $n \geqslant 2$ and $k \geqslant 2$, and easily verified for $k = 1$; hence, for $n \geqslant 2$ and $k \geqslant 1$,

$$a^2 + \frac{(-1)^{n+1}}{a^{2n-2}} > 2a - 1 \geqslant 2a - k = \sqrt{k^2 + 4},$$

whence,

$$(a^{2-n}\sqrt{k^2 + 4})R(k, n) > \sqrt{k^2 + 4};$$

that is,

$$R(k,\ n) > a^{n-2}$$

for $n \geqslant 2$ and $n \geqslant 1$.

To sum up, we have shown that

$$\left(\frac{k + \sqrt{k^2 + 4}}{2}\right)^{n-2} < R(k,\ n) < \left(\frac{k + \sqrt{k^2 + 4}}{2}\right)^{n-1} \tag{3.7}$$

for $n \geqslant 2$ and $k \geqslant 1$. In particular, we have for the Pell numbers P_n,

$$(1 + \sqrt{2})^{n-2} < P_n < (1 + \sqrt{2})^{n-1},$$

and for the Fibonacci numbers F_n,

$$\left(\frac{1 + \sqrt{5}}{2}\right)^{n-2} < F_n < \left(\frac{1 + \sqrt{5}}{2}\right)^{n-1}.$$

4. NUMBER-THEORETIC PROPERTIES OF $R(k, n)$

From the defining recurrence relation (1.1), we obtain

$$
\begin{aligned}
R(k,\ n) &= kR(k,\ n-1) + R(k,\ n-2) \\
&= k(kR(k,\ n-2) + R(k,\ n-3)) + R(k,\ n-2) \\
&= (k^2 + 1)R(k,\ n-2) + kR(k,\ n-3) \\
&= (k^2 + 1)(kR(k,\ n-3) + R(k,\ n-4)) + \\
&\qquad\qquad\qquad + kR(k,\ n-3) \\
&= (k^3 + 2k)R(k,\ n-3) + (k^2 + 1)R(k,\ n-4) \\
&= (k^3 + 2k)(kR(k,\ n-4) + R(k,\ n-5)) + \\
&\qquad\qquad\qquad + (k^2 + 1)R(k,\ n-4) \\
&= (k^4 + 3k^2 + 1)R(k,\ n-4) + (k^3 + 2k)R(k,\ n-5),
\end{aligned}
$$

and so on. Hence,

(i) $R(k,\ 2) = k$,

$R(k,\ 4) = kR(k,\ 3) + R(k,\ 2) = R(k,\ 2)(1 + R(k,\ 3))$,

(continued)

$$R(k, 6) = kR(k, 5) + R(k, 4)$$
$$= R(k, 2)(1 + R(k, 3) + R(k, 5)),$$

. . .

so that, by induction,

$$R(k, 2n) = R(k, 2)\left(\sum_{r=1}^{n} R(k, 2r - 1) \right)$$

$$= k \sum_{r=1}^{n} R(k, 2r - 1),$$

for $n \geqslant 1$, just as we have already obtained [see (2.2)]; we note that, in particular, we have

(a) $F_{2n} = F_2(1 + F_3 + F_5 + \cdots + F_{2n-1})$
$$= F_1 + F_3 + F_5 + \cdots + F_{2n-1}$$

for the Fibonacci numbers,

(b) $P_{2n} = P_2(1 + P_3 + P_5 + \cdots + P_{2n-1})$
$$= 2(P_1 + P_3 + P_5 + \cdots + P_{2n-1}),$$

for the Pell numbers, so that all Pell numbers of the form P_{2n}, $n \geqslant 1$, are even;

(ii) $R(k, 3) = k^2 + 1,$
$$R(k, 6) = (k^2 + 1)R(k, 4) + kR(k, 3)$$
$$= R(k, 3)(k + R(k, 4)),$$
$$R(k, 9) = (k^2 + 1)R(k, 7) + kR(k, 6)$$
$$= R(k, 3)(k^2 + kR(k, 4) + R(k, 7)),$$

. . .

so that, inductively,

$$R(k, 3n) = R(k, 3)\left(\sum_{r=1}^{n} (R(k, 2))^{n-r} R(k, 3r - 2) \right)$$

$$= (k^2 + 1) \sum_{r=1}^{n} k^{n-r} R(k, 3r - 2),$$

for $n \geqslant 1$; in particular, we have

(a) $\quad F_{3n} = F_3(1 + F_4 + F_7 + F_{10} + \cdots + F_{3n-2})$
$$= 2(F_1 + F_4 + F_7 + F_{10} + \cdots + F_{3n-2}),$$

so that all Fibonacci numbers of the form F_{3n}, $n \geqslant 1$, are even,

(b) $\quad P_{3n} = P_3(2^{n-1} + 2^{n-2}P_4 + 2^{n-3}P_7 + \cdots +$
$$+ 2P_{3n-5} + P_{3n-2}),$$
$$= 5 \sum_{r=1}^{n} 2^{n-r}P_{3r-2},$$

so that all Pell numbers of the form P_{3n}, $n \geqslant 1$, are divisible by 5, that is, $P_{3n} \equiv 0 \pmod{5}$;

(iii) $\quad R(k, 4) = k^3 + 2k,$
$$R(k, 8) = (k^3 + 2k)R(k, 5) + (k^2 + 1)R(k, 4)$$
$$= R(k, 4)((1 + k^2) + R(k, 5)),$$
$$R(k, 12) = (k^3 + 2k)R(k, 9) + (k^2 + 1)R(k, 8)$$
$$= R(k, 4)((1 + k^2)^2 + (1 + k^2)R(k, 5) +$$
$$+ R(k, 9)),$$
$$\cdots$$

so that, inductively,

$$R(k, 4n) = R(k, 4)\left(\sum_{r=1}^{n} (R(k, 3)^{n-r}R(k, 4r - 3)) \right)$$

$$= (k^3 + 2k) \sum_{r=1}^{n} (1 + k^2)^{n-r}R(k, 4r - 3),$$

for $n \geqslant 1$; in particular,

(a) $\quad F_{4n} = F_4(2^{n-1} + 2^{n-2}F_5 + 2^{n-3}F_9 + \cdots +$
$$+ 2F_{4n-7} + F_{4n-3}),$$
$$= 3 \sum_{r=1}^{n} 2^{n-r}F_{4r-3},$$

so that all Fibonacci numbers of the form F_{4n}, $n \geqslant 1$, are multiples of 3, that is, $F_{4n} \equiv 0 \pmod{3}$,

(b) $\quad P_{4n} = P_4(5^{n-1} + 5^{n-2}P_5 + 5^{n-3}P_9 + \cdots +$
$$+ 5P_{4n-7} + P_{4n-3})$$

$$= 12 \sum_{r=1}^{n} 5^{n-r}P_{4r-3},$$

so that $P_{4n} \equiv 0 \pmod{12}$ for $n \geqslant 1$.

Proceeding in exactly the same way as in (i)-(iii) above, we can show that, for $p \geqslant 2$,

$$R(k, pn)$$

$$= R(k, p) \sum_{r=1}^{n} (R(k, p-1)^{n-r} R(k, pr-(p-1)) \qquad (4.1)$$

with

$$F_{pn} = F_p \sum_{r=1}^{n} F_{p-1}^{n-r} F_{pr-(p-1)}, \quad p \geqslant 2,$$

and

$$P_{sn} = P_s \sum_{r=1}^{n} P_{s-1}^{n-r} P_{sr-(s-1)}, \quad s \geqslant 2,$$

as the corresponding results for the Fibonacci and Pell numbers, respectively. Clearly, for $n \geqslant 1$,

$$R(k, p) \mid R(k, pn)$$

and, in particular, for Fibonacci and Pell numbers,

$$F_r \mid F_{nr} \quad \text{and} \quad P_r \mid P_{nr}.$$

Now with

$$a = \frac{k + \sqrt{k^2 + 4}}{2} \quad \text{and} \quad b = \frac{k - \sqrt{k^2 + 4}}{2}$$

so that

$$a + b = k, \quad a - b = \sqrt{k^2 + 4}$$
$$1 + ab = 0,$$

the relation (3.3) becomes

$$R(k, n) = \frac{1}{\sqrt{k^2 + 4}}(a^n - b^n) = \frac{a^n - b^n}{a - b},$$

whence,

$$R^2(k, n) + R^2(k, n - 1)$$

$$= \frac{1}{(a - b)^2}[(a^n - b^n)^2 + (a^{n-1} - b^{n-1})^2]$$

$$= \frac{1}{(a - b)^2}[a^{2n-2}(1 + a^2) + b^{2n-2}(1 + b^2)]$$

$$= \frac{a^{2n-1} - b^{2n-1}}{a - b}$$

since, as $1 + ab = 0$, we have $1 + a^2 = a^2 - ab = a(a - b)$ and $1 + b^2 = b^2 - ab = -b(a - b)$.

Thus,

$$R^2(k, n) + R^2(k, n - 1) = R(k, 2n - 1). \qquad (4.2)$$

In particular, with $k = 1$ and $k = 2$, respectively, we have the following known relations for the Fibonacci and Pell numbers:

$$F_n^2 + F_{n-1}^2 = F_{2n-1} \quad \text{and} \quad P_n^2 + P_{n-1}^2 = P_{2n-1}.$$

Next, since

$$R^2(k, n+1) - R^2(k, n-1)$$

$$= (R^2(k, n+1) + R^2(k, n+1)) - (R^2(k, n) + R^2(k, n-1)),$$

we have, by (4.2),

$$R^2(k, n+1) - R^2(k, n-1) = R(k, 2n+1) - R(k, 2n-1)$$

$$= kR(k, 2n), \quad \text{by (1.1)}.$$

Thus,

$$R^2(k, n+1) - R^2(k, n-1) = kR(k, 2n) \qquad (4.3)$$

with

$$F_{n+1}^2 - F_{n-1}^2 = F_{2n} \quad \text{and} \quad P_{n+1}^2 - P_{n-1}^2 = 2P_{2n}$$

as the corresponding relations for the Fibonacci numbers (known, and due to Lucas and Catalan) and Pell numbers, respectively.

Recall that we have [see (2.7)]

$$k(R^2(k, 1) + R^2(k, 2) + \cdots + R^2(k, n))$$
$$= R(k, n)R(k, n+1),$$

and, similarly,

$$k(R^2(k, 1) + R^2(k, 2) + \cdots + R^2(k, n-2))$$
$$= R(k, n-2)R(k, n-1),$$

which, on subtracting, yield

$$k(R^2(k, n) + R^2(k, n-1)$$
$$= R(k, n)R(k, n+1) = R(k, n-2)R(k, n-1),$$

so that, by (4.2),

$$kR(k, 2n-1) = R(k, n)R(k, n+1) -$$
$$- R(k, n-2)R(k, n-1) \qquad (4.4)$$

with

$$F_{2n-1} = F_n F_{n+1} - F_{n-2}F_{n-1}$$

and

$$2P_{2n-1} = P_n P_{n+1} - P_{n-2}P_{n-1}$$

as corresponding relations for the Fibonacci numbers (known, and due to Lucas) and Pell numbers, respectively.

Once again, from the defining recurrence relation (1.1), we have

$$R(k, n+1) = kR(k, n) + R(k, n-1)$$
$$= R(k, n-1)R(k, 1) + R(k, n)R(k, 2).$$

Assume that

$$R(k, n+m)$$
$$= R(k, n-1)R(k, m) + R(k, n)R(k, m+1)$$

holds for some integer $m > 1$. Then

$$R(k, n + (m+1))$$
$$= kR(k, n+m) + R(k, n+m-1)$$
$$= k(R(k, n-1)R(k, m) + R(k, n)R(k, m+1)) +$$
$$+ R(k, n-1)R(k, m-1) + R(k, n)R(k, m)$$

(continued)

$$= R(k, n-1)(kR(k, m) + R(k, m-1)) +$$
$$+ R(k, n)(kR(k, m+1) + R(k, m))$$
$$= R(k, n-1)R(k, m+1) + R(k, n)R(k, m+2).$$

It follows by induction that

$$R(k, n+m) = R(k, n-1)R(k, m) + R(k, n)R(k, m+1)$$

$$(4.5)$$

for integer $m \geqslant 1$ or, with the roles of n and m reversed,

$$R(k, m+n) = R(k, m)R(k, n+1) + R(k, m-1)R(k, n)$$

$$(4.6)$$

for integer $n \geqslant 1$.

In particular, we have

$$F_{n+m} = F_{n-1}F_m + F_n F_{m+1}$$

and

$$P_{n+m} = P_{n-1}P_m + P_n P_{m+1}$$

as corresponding relations for the Fibonacci and Pell numbers, respectively.

Finally, using (4.5), we have

$$R(k, n+m) = R(k, n-1)R(k, m) + R(k, n)R(k, m+1)$$
$$> R(k, m+1)R(k, n), \qquad (4.7)$$

so that, clearly,

$$R(k, n+m) > R(k, n)R(k, m) \qquad (4.8)$$

with

$$F_{n+m} > F_n F_m \quad \text{and} \quad P_{n+m} > P_n P_m$$

as corresponding results for the Fibonacci and Pell numbers, respectively.

Furthermore,

$$R(k, nm) = R(k, n + (m-1)n)$$
$$> R(k, n)R(k, n(m-1))$$
$$> R(k, n)(R(k, n)R(k, n(m-2)))$$

(continued)

$$= R^2(k, n)R(k, n(m-2))$$
$$> R^2(k, n)(R(k, n)R(k, n(m-3)))$$
$$= R^3(k, n)R(k, n(m-3))$$
$$\vdots$$
$$> R^{m-1}(k, n)R(k, n(m-(m-1)))$$
$$= R^m(k, n)$$

through iteration of (4.8). Thus,

$$R(k, nm) > R^m(k, n) \qquad\qquad (4.9)$$

with

$$F_{nm} > F_n^m \quad \text{and} \quad P_{nm} > P_n^m$$

as the corresponding relations for the Fibonacci and Pell numbers.

5. REGION OF CONVERGENCE FOR $R_k(t)$

Although, as an ordinary generating function for the numbers $R(k, n)$, the series

$$R_k(t) = \sum_{n=0}^{\infty} R(k, n)t^n$$

is usually considered as a formal power series, it would still be of some interest to determine its region of convergence.

Making use of the defining recurrence relation (1.1) once again, we have

$$kR(k, n)R(k, n+2) - kR^2(k, n+1)$$
$$= (R(k, n+1) - R(k, n-1))(kR(k, n+1) + R(k, n)) -$$
$$\qquad\qquad - kR^2(k, n+1)$$
$$= -kR(k, n-1)R(k, n+1) - R(k, n-1)R(k, n) +$$
$$\qquad\qquad + R(k, n+1)R(k, n). \qquad\qquad (5.1)$$

Let

$$S(k, n) = kR(k, n-1)R(k, n+1) - R(k, n)R(k, n+1) +$$
$$\qquad\qquad + R(k, n-1)R(k, n).$$

Then, by (1.1),

$$S(k, n) = kR(k, n-1)(kR(k, n) + R(k, n-1)) -$$
$$- (kR(k, n-1) + R(k, n-2))(kR(k, n) +$$
$$+ R(k, n-1)) + R(k, n-1)R(k, n)$$
$$= -kR(k, n-2)R(k, n) - R(k, n-2)R(k, n-1) +$$
$$+ R(k, n)R(k, n-1),$$

so that, by (5.1),

$$S(k, n) = (-1)S(k, n-1). \tag{5.2}$$

Iteration of (5.2) yields

$$S(k, n) = (-1)^{n-1}S(k, 1),$$

whence, by (5.1),

$$S(k, n) = (-1)^n k. \tag{5.3}$$

It follows from (5.1) and (5.3) that

$$R(k, n)R(k, n+2) - R^2(k, n+1) = (-1)^{n+1}, \tag{5.4}$$

which shows that $R(k, n)$ and $R(k, n+1)$ are coprimes, that is,

$$(R(k, n), R(k, n+1)) = 1.$$

In particular, we have:

(i) $F_n F_{n+2} - F_{n+1}^2 = (-1)^{n+1}$,

   so that $(F_n, F_{n+1}) = 1$ for Fibonacci numbers;

(ii) $P_n P_{n+2} - P_{n+1}^2 = (-1)^{n+1}$,

   so that $(P_n, P_{n+1}) = 1$ for Pell numbers.

Now, by (1.1) once again,

$$kR(k, n) < R(k, n+1) = kR(k, n) + R(k, n-1)$$
$$< 2kR(k, n),$$

so that

$$1 < \frac{R(k, n + 1)}{kR(k, n)} < 2. \tag{5.5}$$

Next,

$$\frac{R(k, n + 1)}{kR(k, n)} - \frac{R(k, n + 2)}{kR(k, n + 1)}$$

$$= \frac{R^2(k, n + 1) - R(k, n)R(k, n + 2)}{kR(k, n)R(k, n + 1)},$$

so that, by (5.4),

$$\frac{R(k, n + 1)}{kR(k, n)} - \frac{R(k, n + 2)}{kR(k, n + 1)} = \frac{(-1)^n}{kR(k, n)R(k, n + 1)}.$$

It follows that

$$\lim_{n \to \infty}\left(\frac{R(k, n+1)}{kR(k, n)} - 1\right) = \lim_{n \to \infty} \frac{R(k, n+1) - kR(k, n)}{kR(k, n)}$$

$$= \lim_{n \to \infty} \frac{R(k, n-1)}{kR(k, n)}$$

$$= \lim_{n \to \infty} \frac{R(k, n)}{kR(k, n+1)}$$

that is,

$$\left(\lim_{n \to \infty} \frac{R(k, n+1)}{kR(k, n)}\right) - 1 = \frac{1}{k^2} \lim_{n \to \infty} \frac{1}{\dfrac{R(k, n+1)}{R(k, n)}}. \tag{5.6}$$

Let

$$\alpha = \lim_{n \to \infty} \frac{R(k, n+1)}{kR(k, n)},$$

which exists, by (5.5). Then, necessarily, $\alpha > 0$, and by (5.6),

$$\alpha - 1 = \frac{1}{k^2 \alpha},$$

so that

$$k^2 \alpha^2 - k^2 \alpha - 1 = 0,$$

whence

$$\alpha = \frac{k \pm \sqrt{k^2 + 4}}{2k}$$

that is, since $\alpha > 0$,

$$\alpha = \frac{k + \sqrt{k^2 + 4}}{2k} = \frac{a}{k},$$

where a is as defined in Section 4 above.
    Consequently,

$$\lim_{n \to \infty} \left| \frac{R(k, n + 1) t^{n+1}}{R(k, n) t^n} \right| = a|t| < 1$$

if $|t| < a^{-1} = \dfrac{-k + \sqrt{k^2 + 4}}{2}$.

    In particular,

(i) $F(t) = \displaystyle\sum_{n=0}^{\infty} F_n t^n < \infty$ for $|t| < \dfrac{-1 + \sqrt{5}}{2}$,

(ii) $P(t) = \displaystyle\sum_{n=0}^{\infty} P_n t^n < \infty$ for $|t| < \sqrt{2} - 1$.

6. NUMBERS OF LUCAS TYPE AND FURTHER PROPERTIES OF $R(k, n)$

To help him study the Fibonacci numbers, Lucas introduced the following sequence

$$2, 1, 3, 4, 7, 11, 18, 29, 47, 76, 123, \ldots, \qquad (6.1)$$

which satisfies the recurrence relation

$$\begin{cases} L_n = L_{n-1} + L_{n+1}, & n \geqslant 2, \\ L_0 = 2, \quad L_1 = 1. \end{cases}$$

If we compare the Lucas sequence (6.1) with the Fibonacci sequence

$$0, 1, 1, 2, 3, 5, 8, 13, 21, 34, 55, \ldots,$$

then it is apparent that

$$\begin{cases} L_n = F_{n-1} + F_{n-2}, & n \geqslant 1, \\ L_0 = 2, \end{cases}$$

a result which can be readily proved by setting

$$G_n = F_{n-1} + F_{n+1}, \quad n \geqslant 1,$$

and deriving from the recurrence relation (1.2) for the Fibonacci sequence that

$$\begin{cases} G_n = G_{n-1} + G_{n-2}, & n \geqslant 2, \\ G_0 = 2, \ G_1 = 1. \end{cases}$$

Let us define, for $k \geqslant 1$,

$$L(k, n) = R(k, n-1) + R(k, n+1), \quad n \geqslant 1. \qquad (6.2)$$

Then, by (1.1),

$$\begin{aligned} L(k, n) &= (kR(k, n-2) + R(k, n-3)) + \\ &\quad + (kR(k, n) + R(k, n-1)) \\ &= k(R(k, n-2) + R(k, n)) + \\ &\quad + (R(k, n-3) + R(k, n-1)), \end{aligned}$$

that is, by (6.2),

$$L(k, n) = kL(k, n-1) + L(k, n-2), \ n \geqslant 2. \qquad (6.3)$$

Now, by (6.2), we have

$$L(k, 1) = R(k, 0) + R(k, 2) = k,$$

whence, by (6.3),

$$L(k, 2) = k^2 + L(k, 0).$$

But, by (6.2),

$$\begin{aligned} L(k, 2) &= R(k, 1) + R(k, 3) \\ &= 1 + (k^2 + 1) = 2 + k^2. \end{aligned}$$

Consequently, our numbers $L(k, n)$ satisfy the following recurrence relation for $k \geqslant 1$:

$$\left. \begin{aligned} &L(k, n) = kL(k, n-1) + L(k, n-2), \ n \geqslant 2, \\ &L(k, 0) = 2, \quad L(k, 1) = k. \end{aligned} \right\} \qquad (6.4)$$

As before (see Section 4) with

$$R(k, n) = \frac{a^n - b^n}{a - b},$$

where

$$a = \frac{k + \sqrt{k^2 + 4}}{2} \quad \text{and} \quad b = \frac{k - \sqrt{k^2 + 4}}{2},$$

we can write (6.2) in the form

$$L(k, n) = \frac{1}{a - b}[a^{n-1}(1 + a^2) - b^{n-1}(1 + b^2)],$$

that is, in the form

$$L(k, n) = a^n + b^n,$$

or

$$L(k, n) = \frac{1}{2^{n-1}} \sum_{r \geq 0} \binom{n}{2r} k^{n-2r} (k^2 + 4)^r,$$

so that, in particular,

$$L_n = \frac{1}{2^{n-1}} \sum_{r=0}^{[n/2]} \binom{n}{2r} 5^r.$$

Next, from

$$L^2(k, n) = a^{2n} + 2a^n b^n + b^{2n}$$

and

$$(k^2 + 4)R^2(k, n) = a^{2n} - 2a^n b^n + b^{2n},$$

we obtain

$$L^2(k, n) - (k^2 + 4)R^2(k, n) = 4a^n b^n,$$

that is,

$$L^2(k, n) - (k^2 + 4)R^2(k, n) = 4(-1)^n, \qquad (6.5)$$

which shows that

(i) for k even, $(k^2 + 4)R^2(k, n)$ is always even, so that $L^2(k, n)$ is also even; consequently, $L(k, n)$ is an even number for all n, if k is even;

(ii) for k odd,

   (a) either both $L(k, n)$ and $R(k, n)$ are odd together, and are also coprimes, that is

$$(L(k, n), R(k, n)) = 1 \text{ for all } n,$$

(b) or both $L(k, n)$ and $R(k, n)$ are even together, in which case,

$$(L(k, n), R(k, n)) = 2 \text{ for all } n.$$

The results corresponding to the case $k = 1$, i.e., for the Fibonacci and Lucas numbers are known (see [5], p. 149). Recall that

$$L^2(k, n) = a^{2n} + 2a^n b^n + b^{2n}.$$

Since

$$L(k, n-1)L(k, n+1) = (a^{n-1} + b^{n-1})(a^{n+1} + b^{n+1})$$
$$= a^{2n} + a^{n-1}b^{n+1} +$$
$$+ a^{n+1}b^{n-1} + b^{2n},$$

we have

$$L^2(k, n) - L(k, n-1)L(k, n+1)$$
$$= a^n b^n (2 - a^{-1}b - ab^{-1}) = (-1)^n (2 + a^2 + b^2),$$

that is,

$$L^2(k, n) - L(k, n-1)L(k, n+1) = (-1)^n (k^2 + 4). \tag{6.6}$$

Now, $k^2 + 4$ is a prime number if k is odd, since if d is a divisor of $k^2 + 4$ other than $k^2 + 4$ itself, then $d = 1$, as d must divide both 4 and the odd number k^2. It follows from (6.6) that if k is odd, then $L(k, n)$ and $L(k, n + 1)$ are coprimes, that is,

$$(L(k, n), L(k, n+1)) = 1$$

for all n, provided k is odd.

With the help of the numbers $L(k, n)$, additional divisibility properties of the numbers $R(k, n)$ can now be proved.

For integers m and n, let $d = (m, n)$. Then, there exist integers r and s such that $rm + sn = d$. Since, by (4.6) and (4.5), respectively,

$$R(k, n+m) = R(k, m)R(k, n+1) + R(k, m-1)R(k, n)$$

and

$$R(k, n+m) = R(k, m)R(k, n-1) + R(k, n)R(k, m+1);$$

hence,

$$2R(k,\ n+m) = R(k,\ m)[R(k,\ n+1) + R(k,\ n-1)] +$$
$$+ R(k,\ n)[R(k,\ m+1) + R(k,\ m-1)]$$
$$= R(k,\ m)L(k,\ n) + R(k,\ n)L(k,\ m),$$

and we have

$$2R(k,\ d) = 2R(k,\ rm + sn)$$
$$= R(k,\ rm)L(k,\ sn) + R(k,\ sn)L(k,\ rm).$$

If $(R(k,\ m),\ R(k,\ n)) = h$, then it follows that:

(i) from $h\,|\,R(k,\ m)$ and $R(k,\ m)\,|\,R(k,\ rm)$, we also have $h\,|\,R(k,\ rm)$;

(ii) and similarly, $h\,|\,R(k,\ sn)$.

Consequently, for all k,

$$h\,|\,2R(k,\ d).$$

Now,

(a)   if h is odd, then $h\,|\,R(k,\ d)$;

(b)   if h is even, then both $R(k,\ rm)$ and $R(k,\ sn)$ are even, so that, by (6.5), all the numbers $R(k,\ rm)$, $R(k,\ sn)$, $L(k,\ rm)$, and $L(k,\ sn)$ are even, and we can write

$$R(k,\ d) = R(k,\ rm)(\tfrac{1}{2}L(k,\ sn)) + R(k,\ sn)(\tfrac{1}{2}L(k,\ rm))$$

and see that $h\,|\,R(k,\ d)$.

Next, since $(m,\ n) = d$, we have

$$m = q_1 d \quad \text{for some integer } q_1$$
and
$$n = q_2 d \quad \text{for some integer } q_2,$$

so that

$$R(k,\ d)\,|\,R(k,\ m)$$
and
$$R(k,\ d)\,|\,R(k,\ n),$$

that is,

$$R(k, d) \mid (R(k, m), R(k, n)) = h.$$

Hence,

$$h = R(k, d),$$

that is,

$$(R(k, m), R(k, n)) = R(k, d) \qquad\qquad (6.7)$$

where $d = (m, n)$.

In particular, if m and n are coprimes, that is, $d = 1$, then:

(i) $R(k, m)$ and $R(k, n)$ are also coprimes, since, by (6.7),

$$(R(k, m), R(k, n)) = R(k, 1) = 1;$$

(ii) from the fact that $R(k, m) \mid R(k, mn)$ and $R(k, n) \mid R(k, mn)$ together with the result just obtained in (i), we have

$$R(k, m)R(k, n) \mid R(k, mn).$$

The special cases

$$(P_m, P_n) = 1 \text{ and } P_m P_n \mid P_{mn} \text{ if } (m, n) = 1$$

and

$$(F_m, F_n) = 1 \text{ and } F_m F_n \mid F_{mn} \text{ if } (m, n) = 1$$

are known results for the Pell and Fibonacci numbers, respectively.

7. CONCLUDING REMARKS

By way of conclusion, we note that if we write our defining recurrence relation

$$\left.\begin{array}{l} R(k, n) = kR(k, n-1) + R(k, n-2), \quad n \geqslant 2 \\ R(k, 0) = 0, \quad R(k, 1) = 1, \end{array}\right\}$$

in the form

$$\left.\begin{array}{l} r_n(x) = xr_{n-1}(x) + r_{n-2}(x), \quad n \geqslant 2 \\ r_0(x) = 0, \quad r_1(x) = 1, \end{array}\right\} \qquad (7.1)$$

whose solution is, by (3.7), the polynomial

$$r_n(x) = \sum_{p=0}^{[(n-1)/2]} \left. \binom{n-p-1}{p} x^{n-p-1}, \right\}$$ (7.2)

then we readily see that (7.2) is the Fibonacci polynomial. However, considered as numbers, the $R(k, n)$'s lend themselves more easily to algebraic manipulation which, with the help of known, and elementary, techniques of combinatorics and number theory, has given us a wealth of information on their properties as well as on the properties of the Fibonacci numbers F_n and the Pell numbers P_n.

Riordan, in [8], considered a polynomial similar to (7.2), which he denoted

$$u_n(x) = \sum_{p=0}^{[n/2]} \binom{n-p}{p} x^p,$$ (7.3)

because of its association with Chebyshev relations

$$\begin{cases} U_n(x) = \dfrac{\sin(n+1)\theta}{\sin \theta}, \\ \cos \theta = x. \end{cases}$$

The polynomial (7.3) can be shown to satisfy the recurrence relation

$$\begin{cases} u_n(x) = u_{n-1}(x) + x u_{n-2}(x), \quad n \geqslant 2 \\ u_0(x) = 1, \quad u_1(x) = 1 \end{cases}$$

whose iteration would lead to

$$u_{n+m}(x) = u_m(x) u_n(x) + x u_{m-1}(x) u_{n-1}(x)$$

from which the combinatorial identity

$$\binom{n+m-p}{p} = \sum_{r \geqslant 0} \left[\binom{n-r}{r}\binom{m-p+r}{p-r} + \right.$$

$$\left. + \binom{n-1-r}{r}\binom{m-p+r}{p-1-r} \right]$$

can be derived. Clearly, similar identities can be derived from various relations we have obtain for our numbers $R(k, n)$.

Although the Fibonacci numbers F_n and the Pell numbers P_n are special cases of our numbers $R(k, n)$, instead of giving the latter the label of generalized Fibonacci or Pell numbers, we are of the opinion that it would be more appropriate to think of them as members of a class of numbers to which both of the numbers F_n and P_n also belong.

REFERENCES

[1]   Bong, N. H.   'On Figurate Numbers That Can Be Repre-
      sented by Truncated Tetrahedra.' *Menemui Matematik* **6**,
      no. 1 (1984).

[2]   Burton, D. M.   *Elementary Number Theory*.   Boston: Allyn
      & Bacon, 1980.

[3]   Cohen, D. I. A.   *Basic Techniques of Combinatorial
      Theory*.   New York: John Wiley & Sons, 1978.

[4]   Entringer, R. C., and Slater, P. J.   'Gossips and Tele-
      graphs.' *J. of the Franklin Institute* **307**, no. 6 (June
      1979):353-359.

[5]   Hardy, G. H., and Wright, E. M.   *An Introduction to the
      Theory of Numbers*.   4th ed.   London: OUP, 1962.

[6]   Hoggatt, V. E., Jr., and Bicknell-Johnson, Marjorie.
      'Reflections Across Two and Three Glass Plates.' *The
      Fibonacci Quarterly* **17**, no. 2 (1979):118-141.

[7]   Jordan, C.   *Calculus of Finite Differences*.   New York:
      Chelsea, 1960.

[8]   Riordan, J.   *Combinatorial Identities*.   New York: John
      Wiley & Sons, 1968.

Herta T. Freitag

A PROPERTY OF UNIT DIGITS OF
FIBONACCI NUMBERS

A computer printout which came to my attention listed the
first Fibonacci numbers horizontally such that five columns
occurred. Reading vertically, any column gave the sequence

$$\{F_{n+5k}\}, \quad n \in \{1, 2, 3, 4, 5\}, \quad k \in \{0, 1, 2, 3, \ldots\}.$$

(The third column, for instance, gave F_3, F_8, F_{13}, $\ldots$.)
It appeared that the unit digit of the sum of two successive
entries equalled the unit digit of the third entry.
    This triggered three questions:

(a) Is this relationship correct for all n, i.e.,

$$F_n + F_{n+5} \overset{?}{\equiv} F_{n+10} \ (\text{mod } 10)$$

(1)

   for n being a nonnegative integer?

(b) Are there any x-values other than 1 or 5 such that

$$F_n + F_{n+x} \equiv F_{n+2x} \ (\text{mod } 10)$$

(2)

holds?

(c) If such x-values exist, would they also be applicable
     to more general recursive sequences?

     Using the formula

$$F_{n+t} = F_t F_{n+1} + F_{t-1} F_n$$

(which is readily obtainable by induction on t), relationship (2) becomes

$$(1 + F_{x-1})F_n + F_x F_{n+1} \equiv F_{2x-1}F_n + F_{2x}F_{n+1} \ (\text{mod } 10).$$

(3)

A. N. Philippou et al. (eds.), Fibonacci Numbers and Their Applications, 39–41.

Re. (a)

It is easily seen that (3) holds for $x = 5$, as

$$1 + F_4 \equiv F_9 \pmod{10}; \quad F_5 \equiv F_{10} \pmod{10}.$$

Thus, relationship (1) is satisfied for all n.

Re. (b)

Now we wish to determine S, the set of all x-values for which relationship (2) is valid.

First, we note that

$$F_{p+60q} \equiv F_p \pmod{10},$$

where p and q are nonnegative integers. Thus, restriction to $A = \{x : 1 < x \leqslant 59\}$ suffices.

By inspection, we conjecture that the only subset of A that satisfies relationship (2) becomes

$$\overline{S} = \{x : x \equiv 1 \text{ or } 5 \pmod{12}\}$$

$$= \{1, 5, 13, 17, 25, 29, 37, 49, 53\}$$

$$\left.\right\} \quad (4)$$

such that

$$S = \{y : y \equiv x \pmod{60}\}.$$

The proof of this conjecture utilizes (3) and requires two cases.

Case 1: $x \equiv 1 \pmod{12}$. Letting $x = 12k + 1$, k a positive integer, we have

$$1 + F_{12k} \stackrel{?}{\equiv} F_{24k+1} \pmod{10}$$
$$F_{12k+1} \stackrel{?}{\equiv} F_{24k+2} \pmod{10}$$
$$\left.\right\}. \quad (5)$$

Now

$$F_{12k} \equiv 4K \pmod{10}$$
$$F_{24k+1} \equiv 4K + 1 \pmod{10}$$
$$\left.\right\}.$$

Further,

$$F_{12k+1} \equiv 2K + 1 \equiv F_{24k+2} \pmod{10}.$$

Thus, the questioned conditions (5) are satisfied.

Case 2: $x \equiv 5 \pmod{12}$ is treated analogously such that

$$F_{12k+4} \equiv 4K + 3 \pmod{10},$$

$$F_{24k+9} \equiv 4K + 4 \pmod{10},$$

and

$$F_{12k+5} \equiv 2K + 5 \equiv F_{24k+10} \pmod{10},$$

which, again, satisfies relationship (5).

This establishes conjecture (4), thereby determining the x-values sought in (2).

Re. (c)

It is immediately apparent that the sequence $\{A_n\}_{n=1}^{\infty}$, defined by $A_1 = p$, $A_2 = q$, p and q nonnegative integers, and $A_{n+2} = A_{n+1} + A_n \; \forall_n \in \mathbb{N}$ satisfies the conditions specified.

Katuomi Hirano

SOME PROPERTIES OF THE DISTRIBUTIONS
OF ORDER k

1. INTRODUCTION

In this paper k is a fixed positive integer, and p and
$q = 1 - p$ are real numbers in the interval (0, 1). Let $T_{k,1}$
be a random variable denoting the number of trials until the
first occurrence of the kth consecutive success in independent trials with constant success probability p. Philippou
& Muwafi (1982) introduced the Fibonacci sequence of order k
and discussed the relationship between it and the distribution of $T_{k,1}$, which was called the geometric distribution of
order k [to be denoted by $G_k(p)$]. Recently, Philippou,
Georghiou, & Philippou (1983), Philippou (1983; 1984),
Hirano, Kuboki, Aki, & Kuribayashi (1984), and Aki, Kuboki,
& Hirano (1984) discussed several distributions of order k
that are derived from $G_k(p)$. We are interested here in the
distributions of order k with success probability p. These
distributions of order k are defined as follows.

The Negative Binomial Distribution of Order k
[Abbreviated $NB_k(r, p)$]

A random variable X is said to have the negative binomial
distribution of order k with parameters p and r (any positive real number), if

$$
Pr\{X = x\} = \sum_{x_1, \ldots, x_k} \binom{x_1 + \cdots + x_k + r - 1}{x_1, \ldots, x_k, r - 1} \times
$$

$$
\times \, p^{x + kr - [kr]} \left(\frac{q}{p}\right)^{x_1 + \cdots + x_k},
$$

$$
x = [kr], [kr] + 1, \ldots, \qquad (1.1)
$$

where the summation is over all nonnegative integers x_1,
$\ldots, x_k$ such that $x_1 + 2x_2 + \cdots + kx_k = x - [kr]$. It is
observed that $NB_k(1, p)$ is $G_k(p)$.

43

A. N. Philippou et al. (eds.), Fibonacci Numbers and Their Applications, 43–53.

The Poisson Distribution of Order k
[Abbreviated $P_k(\lambda)$]

A random variable X is said to have the Poisson distribution of order k with parameter λ $(\lambda > 0)$ if

$$Pr\{X = x\} = \sum_{x_1, \ldots, x_k} e^{-k\lambda} \frac{\lambda^{x_1 + \cdots + x_k}}{x_1! \ldots x_k!},$$

$$x = 0, 1, 2, \ldots, \quad (1.2)$$

where the summation is over all nonnegative integers x_1, $\ldots$, x_k such that $x_1 + 2x_2 + \cdots + kx_k = x$.

The Logarithmic Series Distribution of Order k
[Abbreviated $LS_k(p)$]

A random variable X is said to have the logarithmic series distribution of order k with parameter p if

$$Pr\{X = x\} = \sum_{x_1, \ldots, x_k} \frac{(x_1 + \cdots + x_k - 1)!}{(-k \log p)x_1! \ldots x_k!} \times$$

$$\times p^x \left(\frac{q}{p}\right)^{x_1 + \cdots + x_k},$$

$$x = 1, 2, 3, \ldots, \quad (1.3)$$

where the summation is over all nonnegative integers x_1, $\ldots$, x_k such that $x_1 + 2x_2 + \cdots + kx_k = x$.

Next we shall remark on the generalization of discrete distributions. Let $\psi_1(t)$ and $\psi_2(t)$ be the probability generating functions (abbr. pgf) of two discrete distributions F_1 and F_2, respectively. Then the resulting function, $\psi_1(\psi_2(t))$, is also a pgf. The distribution whose pgf is $\psi_1(\psi_2(t))$ is called a generalized F_1 distribution (generalized) by the generalizer F_2. We denote it by the symbolic form $F_1 \vee F_2$.

Now let α be a positive real number that satisfies $\psi_1(\alpha) < \infty$. Then the distribution with pgf $\psi_1(\alpha\psi_2(t))/\psi_1(\alpha)$ is called the α-generalized F_1 distribution by the generalizer F_2. When $\alpha = 1$, the α-generalization is the same as the usual generalization stated above.

In Section 2 we introduce the binomial distribution of order k and study some of its properties. In Section 3 we discuss the relationship between $NB_k(r, p)$ and $LS_k(p)$. In the last section we show that the probability of a class of α_0-generalized distributions, given by Aki, Kuboki, & Hirano (1984), is expressed as an explicit form. Some new distributions of order k belonging to the class are also given.

2. BINOMIAL DISTRIBUTION OF ORDER k

In this section we introduce the binomial distribution of order k and examine some of its properties.

Let $N_{k,n}$ be a random variable denoting the number of occurrences of the kth consecutive success in n independent trials with success probability p. The distribution of $N_{k,n}$ will be referred to as the binomial distribution of order k [denoted by $B_k(n, p)$]. Let $T_{k,r}$ be a random variable denoting the number of trials until the rth occurrence of the kth consecutive success in the trials. Then $T_{k,r}$ is distributed as $NB_k(r, p)$. Feller (1968) notes that the relationship between the probability distributions of $T_{k,r}$ and $N_{k,n}$ is given by the identity

$$Pr\{N_{k,n} \geqslant r\} = Pr\{T_{k,r} \leqslant n\}, \quad r = 0, 1, \ldots, \left[\frac{n}{k}\right]$$

(2.1)

and discusses the asymptotic behavior of the distributions of $N_{k,n}$ and $T_{k,r}$ for large n and r, respectively. We shall investigate the exact behavior.

We can use (2.1) to obtain the pgf and the νth moment of the binomial distribution of order k. In fact, we have the following proposition, which will be easily checked.

Proposition 2.1: Let X be a random variable distributed as $B_k(n, p)$. Then its pgf, $\psi_B(t)$, and ν^{th} moment, $E(X^\nu)$, are given by

$$\psi_B(t) = 1 + (t - 1) \sum_{i=1}^{[n/k]} t^{i-1} a(i),$$

(2.2)

and

$$E(X^\nu) = \sum_{i=1}^{[n/k]} \{i^\nu - (i - 1)^\nu\} a(i),$$

(2.3)

respectively, where $a(i) = Pr\{T_{k,i} \leqslant n\}$.

The table below gives some means of $B_k(n, p)$.

k \ n	4	5	6
2	$p^2(3 - 2p + p^2)$	$p^2(4 - 3p + 2p^2 - p^3)$	$p^2(5 - 4p + 3p^2 - 2p^3 + p^4)$
3	$p^3(2 - p)$	$p^3(3 - 2p)$	$p^3(4 - 3p + p^3)$
4	p^4	$p^4(2 - p)$	$p^4(3 - 2p)$
5		p^5	$p^5(2 - p)$

$$E(N_{n,n}) = p^n, \quad (n \geqslant 1);$$

$$E(N_{n-1,n}) = p^{n-1}(1 + q), \quad (n \geqslant 3);$$

$$E(N_{n-2,n}) = p^{n-2}(1 + 2q), \quad (n \geqslant 5);$$

$$E(N_{n-3,n}) = p^{n-3}(1 + 3q), \quad (n \geqslant 7).$$

Now we define $a(0) = 1$ and $a([n/k] + 1) = 0$. Then, from (2.1),

$$Pr\{N_{k,n} = x\} = a(x) - a(x + 1),$$

$$x = 0, 1, 2, \ldots, [n/k]. \quad (2.3)$$

Further, we obtain another representation. The proof is given along the lines of Philippou & Muwafi (1982).

Proposition 2.2:

$$Pr\{N_{k,n} = x\} = \sum_{m=0}^{k-1} \sum_{x_1, \ldots, x_k} \binom{x_1 + \cdots + x_k + x}{x_1, \ldots, x_k, x} \times$$

$$\times p^n \left(\frac{q}{p}\right)^{x_1 + \cdots + x_k}$$

$$x = 0, 1, 2, \ldots, [n/k], \quad (2.4)$$

where the inner summation is over all nonnegative integers $x_1, \ldots, x_k$ such that $x_1 + 2x_2 + \cdots + kx_k = n - m - kx$.

Proof: We refer to the outcome with probability p as "success," S, and to the other as "failure," F. For $j = 1, 2, \ldots, k$, let e_j denote the pattern of the outcomes $\underbrace{S \ldots S}_{j-1}F$, and e denote $\underbrace{S \ldots S}_{k}$. Then, the event "$N_{k,n} = x$" occurs if and only if the outcome of the trial is of the form

$$a_1 a_2 \cdots a_{x_1 + \cdots + x_k + x} \underbrace{S \cdots S}_{m}, \quad (0 \leqslant m \leqslant k - 1).$$

where x_j of the a's are e_j, and where x's should satisfy the relation

$$x_1 + 2x_2 + \cdots + kx_k + kx + m = n, \quad (0 \leqslant m \leqslant k - 1).$$

Fix $x_1, \ldots, x_k$. Then the number of the above arrangements

$$\binom{x_1 + \cdots + x_k + x}{x_1, \ldots, x_k, x},$$

and each one of them has probability

$$P(a_1 a_2 \cdots a_{x_1 + \cdots + x_k + x} \underbrace{S \cdots S}_{m})$$

$$= \{P(e_1)\}^{x_1} \{P(e_2)\}^{x_2} \cdots \{P(e_k)\}^{x_k} \{P(e)\}^k \{P(S)\}^m$$

$$= p^{m + x_2 + 2x_3 + \cdots + (k-1)x_k + kx} q^{x_1 + \cdots + x_k}, \quad (0 \leqslant m \leqslant k - 1),$$

by the independence of the trials, the definition of e_j $(1 \leqslant j \leqslant k)$ and e, and $P(S) = p$ and $P(F) = q$. Hence,

$$P\bigg(\text{all } a_1 a_2 \cdots a_{x_1 + \cdots + x_k + x} \underbrace{S \cdots S}_{m}; \; x_j \text{ fixed } (1 \leqslant j \leqslant k),$$

$$m \text{ fixed } (0 \leqslant m \leqslant k - 1), \text{ and } \bigg(\sum_{j=1}^{k} jx_j\bigg) + kx + m = n\bigg)$$

$$= \binom{x_1 + \cdots + x_k + x}{x_1, \ldots, x_k, x} p^{m + x_2 + 2x_3 + \cdots + (k-1)x_k + kx} q^{x_1 + \cdots + x_k}.$$

But for fixed m $(0 \leqslant m \leqslant k - 1)$ the nonnegative integers $x_1, \ldots, x_k$ may vary, subject to the condition

$$x_1 + 2x_2 + \cdots + kx_k + kx + m = n.$$

Further, the integer m may also vary. Consequently, since all of these events are mutually exclusive, we obtain (2.4).

The binomial distribution of order k has important applications. Chiang & Niu (1981) introduced the consecutive-k-of-n : F system and indicated its relevance to some telecommunication and oil pipeline systems. Bollinger & Salvia (1982) also remarked that such systems frequently arise in the design of integrated circuitry. Thereafter, several articles appeared in *IEEE Transactions on Reliability* which have some relationship to the problem. The consecutive-k-of-$n:F$ system contains n independent components, each with probabilities q and p of operating or failing, and the system fails if k consecutive components fail. We point out that the failure probability of the consecutive-k-of-$n:F$ system is $1 - Pr\{N_{k,n} = 0\}$.

Remark; Proposition 2.2 has also been obtained independently by Philippou & Makri (1984).

3. RELATION BETWEEN $NB_k(r, p)$ AND $LS_k(p)$

Aki, Kuboki, & Hirano (1984) defined $NB_k(r, p)$ for any positive real number r and introduced $LS_k(p)$. Roughly speaking, the latter is a limiting form of the former as $r \to 0$. Let X be a random variable distributed as $NB_k(r, p)$. Then we call the distribution of $Y = X - [kr]$ the shifted negative binomial distribution of order k [denoted by $\overline{NB}_k(r, p)$]. It is clear that $\overline{NB}_k(r, p)$ is the probability distribution on the set $\{0, 1, 2, \ldots\}$ and the pgf, $\psi_{\overline{NB}}(t; k)$, of it is given by

$$\psi_{\overline{NB}}(t; k) = p^{kr}\left\{\frac{1 - pt}{1 - t + qp^k t^{k+1}}\right\}^r. \tag{3.1}$$

Then, we have the following proposition.

Proposition 3.1: Let X be a random variable distributed as $\overline{NB}_k(r, p)$, and assume that $r > 0$. Then $Pr\{X = x | X \geqslant 1\}$ converges to the right-hand side of (1.3) as $r \to 0$.

A similar statement about $NB_k(r, p)$ is given by Aki, Kuboki, & Hirano (1984). The proof is the same as given there.

It is well known that the usual negative binomial distribution $NB_1(\cdot, p)$, is a generalized Poisson distribution by the generalizer of the usual logarithmic series distribution, $LS_1(p)$; that is, $\overline{NB}_1(-\lambda/\log p, p) = P_1(\lambda) \vee LS_1(p)$. We now obtain the corresponding relation.

Proposition 3.2: $\overline{NB}_k(-\lambda/\{k \log p\}, p) = P_1(\lambda) \vee LS_k(p)$.

Proof: Since the pgf's of $P_1(\lambda)$ and $LS_k(p)$ are

$$\psi_P(t; 1) = \exp(-\lambda + \lambda t)$$

and

$$\psi_{LS}(t; k) = \alpha(p)\log(1 - pt)/(1 - t + qp^k t^{k+1}),$$

respectively, where $\alpha(p) = -1/\{k \log p\}$, the pgf of $P_1(\lambda) \vee LS_k(p)$ is

$$\psi_P(\psi_{LS}(t; k); 1).$$

From (3.1), the result is obvious.

4.   A CLASS OF α_0-GENERALIZED DISTRIBUTIONS

In this section we introduce the "k-point" distribution and discuss the relation between it and the distributions of order k. A class of α_0-generalized distributions is introduced and some of its properties are examined.

A random variable Y is said to have the k-point distribution with parameters v_1, v_2, ..., v_k [denoted by $K(v_1, \ldots, v_k)$] if

$$Pr\{Y = j\} = v_j, \quad j = 1, 2, \ldots, k;$$

$$v_j \geqslant 0, \quad v_k \neq 0, \quad \sum_{j=1}^{k} v_j = 1.$$

Let $K(v_1, \ldots, v_k : n)$ be the distribution of the sum of n independent, identically distributed random variables from $K(v_1, \ldots, v_k)$. Then we obtain the following proposition.

Proposition 4.1:

(i) The pgf's of $K(v_1, \ldots, v_k)$ and $K(v_1, \ldots, v_k : n)$ are

$$\psi_K(t) = \sum_{j=1}^{k} v_j t^j \quad \text{and} \quad \{\psi_K(t)\}^n,$$

respectively.

(ii) The mth moment about zero is $\mu_m' = \sum_{j=1}^{k} j^m v_j$.

(iii) The probability function of $K(v_1, \ldots, v_k : n)$ is written as

$$Pr\{X = j\} = \sum_{x_1, \ldots, x_k} \binom{x_1 + \cdots + x_k}{x_1, \ldots, x_k} v_1^{x_1} \ldots v_k^{x_k},$$

$$j = n,\ n + 1,\ \ldots,\ kn,$$

where the summation is over all nonnegative integers x_1, $\ldots$, x_k such that

$$x_1 + 2x_2 + \cdots + kx_k = j$$

and

$$x_1 + x_2 + \cdots + x_k = n.$$

The proof is easily checked.

Charalambides (1977) pointed out that the probabilities of the class of the usual generalized distributions are expressed in terms of the Bell polynomials. We shall give the probability function of the α-generalized F_1 distribution by the generalizer $K(v_1, \ldots, v_k)$.

Suppose that the support of F_1 is $\{0, 1, 2, \ldots\}$ (or $\{1, 2, 3, \ldots\}$). Let $\psi_1(t)$ be a pgf of F_1. Then, we obtain the following proposition.

Proposition 4.2: Let X be a random variable distributed as the α-generalized F_1 distribution by the generalizer $K(v_1, \ldots, v_k)$. Then we obtain

$$Pr\{X = j\}$$

$$= \sum_{x_1, \ldots, x_k} \frac{\alpha^{x_1 + \cdots + x_k} v_1^{x_1} \ldots v_k^{x_k}}{x_1! \ldots x_k!} \psi_1^{(x_1 + \cdots + x_k)}(0) / \psi_1(\alpha),$$

$$j \geqslant 0 \ (\text{resp. } j \geqslant 1), \quad (4.2)$$

where $\psi_1^{(n)}(0) = (d/dt)^n \psi_1(t)|_{t=0}$ and the summation is over all nonnegative integers $x_1, \ldots, x_k$ such that

$$x_1 + 2x_2 + \cdots + kx_k = j.$$

Proof: Notice that $(d/dt)^\ell \psi_K(t)|_{t=0} = \ell! v_\ell$, $\ell = 1, 2, \ldots, k$ and $\psi_K(0) = 0$. Fix j. For $j \leqslant k$, the summation over all nonnegative integer solutions of

$$s_1 + 2s_2 + \cdots + js_j + (j + 1)s_{j+1} + \cdots + ks_k = j$$

is equal to the one of

$$s_1 + 2s_2 + \cdots + js_j = j, \quad s_{j+1} = \cdots = s_k = 0.$$

For $j > k$, the nonzero terms in $(d/dt)^j \psi_1(\alpha \psi_K(t))/\psi_1(\alpha)$ remains only if $s_{k+1} = \cdots = s_j = 0$. Hence, we obtain the formula (4.2).

Let $\mathscr{A}$ be a class of the α_0-generalized F_1 distribution by the generalizer

$$K(1/\alpha_0, \ p/\alpha_0, \ \ldots, \ p^{k-1}/\alpha_0),$$

where $\alpha_0 = (1 - p^k)/(1 - p)$, ($\alpha_0 = k$ for $p = 1$). Then $\mathscr{A}$ contains $\overline{NB}_k(r, p)$, $LS_k(p)$, and $P_k(\lambda)$ [see Aki, Kuboki, & Hirano (1984)]. Note that $B_k(n, p)$ does not belong to $\mathscr{A}$. We now have the corollary to the proposition.

Corollary: For any $F \in \mathscr{A}$, let X be a random variable distributed as F. Then

$$Pr\{X = j\}$$

$$= \sum_{x_1, \ \ldots, \ x_k} \frac{1}{x_1! \ \cdots \ x_k!} p^{j-(x_1+\cdots+x_k)} \psi_1^{(x_1+\cdots+x_k)}(0)/\psi_1(\alpha_0),$$

$$(4.3)$$

where the summation is over all nonnegative integers x_1, $\ldots$, x_k such that $x_1 + 2x_2 + \cdots + kx_k = j$. For $P_k(\lambda)$, (4.3) holds with $p = 1$.

Remark: Consider the case $\psi_1(t) = (pt + q)^n$, which is the pgf of the usual binomial distribution $B_1(n, p)$. Then the distribution given by (4.3) is not $B_k(n, p)$.

We shall now give some distributions of order k that belong to $\mathscr{A}$. Let $P_1(m)$ be the usual Poisson distribution with pgf $\psi_p(t; 1) = \exp(mt - m)$. Then consider the three generalized Poisson distributions $P_1(m) \vee \overline{NB}_k(r, p)$, $P_1(m) \vee LS_k(p)$, and $P_1(m) \vee P_k(\lambda)$. First, it is observed that $P_1(m) \vee \overline{NB}_1(1, p)$ is the usual (shifted) Pólya-Aeppli distribution. We call $P_1(m) \vee \overline{NB}_k(1, p)$ the (shifted) Pólya-Aeppli distribution of order k. The pgf of $P_1(m) \vee \overline{NB}_k(r, p)$ is $\psi_p(\psi_{\overline{NB}}(t; k); 1)$. Next, $P_1(m) \vee LS_k(p)$ is as given in Section 3. This distribution is $\overline{G}_k(p)$ for

$m = -k \log p$. Last, it is well known that $P_1(m) \vee P_1(\lambda)$ is Neyman's Type A distribution. Thus, we call $P_1(m) \vee P_k(\lambda)$ Neyman's Type A distribution of order k. The pgf of $P_1(m) \vee P_k(\lambda)$ is $\psi_p(\psi_p(t; k); 1)$. Therefore, the next proposition will be easily checked.

Proposition 4.3:

(i) $P_1(p^{-(k-1)r}) \vee \overline{NB}_k(r, p)$ and $P_1(k) \vee LS_k(p)$ are obtained by the α_0-generalization of $P_1(1) \vee \overline{NB}_1(r, p)$ and $P_1(1) \vee LS_1(p)$, respectively, by the common generalizer

$$K(1/\alpha_0, p/\alpha_0, \ldots, p^{k-1}/\alpha_0).$$

(ii) $P_1(e^{k\lambda}) \vee P_k(\lambda)$ is obtained by the k-generalization of $P_1(e^\lambda) \vee P_1(\lambda)$ by the generalizer

$$K(1/k, 1/k, \ldots, 1/k).$$

We note that $P_1(p^{-(k-1)r}) \vee \overline{NB}_k(r, p)$, $P_1(k) \vee LS_k(p)$, and $P_1(e^{k\lambda}) \vee P_k(\lambda)$ belong to $\mathscr{A}$, and these probability functions are given by (4.3). The last one is also given by (4.3) with $p = 1$.

ACKNOWLEDGMENTS

The author is thankful to Mr. S. Aki for pointing out the fact that the failure probability of the system in Section 2 is the nonzero probability of the binomial distribution of order k.

REFERENCES

Aki, S.; Kuboki, H.; and Hirano, K. (1984). 'On Discrete Distributions of Order k.' *Ann. Inst. Statist. Math.* 36, A, 431–440.

Bollinger, R. C., and Salvia, A. A. (1982). 'Consecutive-k-out-of-n: F Networks.' *IEEE Transactions on Reliability* R-31, April, 53–55.

Chiang, D., and Niu, S. C. (1981). 'Reliability of Consecutive-k-out-of-n: F System.' *IEEE Transactions on Reliability* R-30, April, 87–89.

Charalambides, Ch. A. (1977). 'On the Generalized Discrete Distributions and the Bell Polynomials.' *Sankhya* **39**, Series B, 36–44.

Feller, W. (1968). *An Introduction to Probability Theory and Its Applications*. Vol. I. 3rd ed. New York: John Wiley & Sons, Inc.

Hirano, K.; Kuboki, H.; Aki, S.; and Kuribayashi, A. (1984). 'Figures of Probability Density Functions in Statistics II—Discrete Univariate Case.' *Computer Science Monographs*, No. 20, The Inst. of Statist. Math.

Philippou, A. N. (1983). 'The Poisson and Compound Poisson Distribution of Order *k* and Some of Their Properties.' *Zapiski Nauchnyka Seminarov LOMI, im. V. A. Steklova AN SSSR* **130**, 175–180 (Russian, English summary).

Philippou, A. N. (1984). 'The Negative Binomial Distribution of Order *k* and Some of Its Properties.' *Biometrical Journal* **26**, 789–794.

Philippou, A. N.; Georghiou, C.; and Philippou, G. N. (1983). 'A Generalized Geometric Distribution and Some of Its Properties.' *Statistics and Probability Letters* **1**, 171–175.

Philippou, A. N., and Makri, F. S. (1984). 'Successes, Runs, and Longest Runs.' Submitted for publication.

Philippou, A. N., and Muwafi, A. A. (1982). 'Waiting for the *k*th Consecutive Success and the Fibonacci Sequence of Order *k*.' *The Fibonacci Quarterly* **20**, no. 1, 28–32.

A. F. Horadam and Br. J. M. Mahon

CONVOLUTIONS FOR PELL POLYNOMIALS

PART I: BACKGROUND

1. Introduction

Given two polynomial sequences $\{a_n(x)\}$ and $\{b_n(x)\}$ where n is an integer ≥ 1, their *first convolution sequence* $\{c_n(x)\}$ is defined by

$$c_n(x) = \sum_{j=1}^{n} a_j(x) b_{n+1-j}(x) = \sum_{j=1}^{n} b_j(x) a_{n+1-j}(x).$$

$$(1.1)$$

When $b_n(x) = a_n(x)$, the *first convolution sequence* $\{d_n(x)\}$ of $\{a_n(x)\}$ is given by

$$d_n(x) = \sum_{j=1}^{n} a_j(x) a_{n+1-j}(x).$$

$$(1.2)$$

The *Pell polynomials* $P_n(x)$ and *Pell-Lucas polynomials* $Q_n(x)$ are defined by the recurrence relations

$$P_{n+2}(x) = 2xP_{n+1}(x) + P_n(x), \quad \begin{aligned} P_0(x) &= 0, \\ P_1(x) &= 1, \end{aligned} \quad (1.3)$$

and

$$Q_{n+2}(x) = 2xQ_{n+1}(x) + Q_n(x), \quad \begin{aligned} Q_0(x) &= 2, \\ Q_1(x) &= 2x, \end{aligned} \quad (1.4)$$

respectively.

In this paper, we investigate the properties of the convolution sequences for $\{P_n(x)\}$ and $\{Q_n(x)\}$, i.e., we relate (1.1) and (1.2) to (1.3) and (1.4).

Generating functions for these polynomial sequences $\{P_n(x)\}$ and $\{Q_n(x)\}$ are, respectively,

$$(1 - 2xy - y^2)^{-1} = \sum_{r=0}^{\infty} P_{r+1}(x) y^r = \sum_{r=1}^{\infty} P_r(x) y^{r-1}$$

$$(1.5)$$

A. N. Philippou et al. (eds.), Fibonacci Numbers and Their Applications, 55–80.
© *1986 by D. Reidel Publishing Company.*

and

$$(2x + 2y)(1 - 2xy - y^2)^{-1} = \sum_{r=0}^{\infty} P_{r+1}(x)y^r. \qquad (1.6)$$

Explicit summation formulas are, respectively,

$$P_n(x) = \sum_{m=0}^{[(n-1)/2]} \binom{n-m-1}{m}(2x)^{n-2m-1} \qquad (1.7)$$

and

$$Q_n(x) = \sum_{m=0}^{[n/2]} \frac{n}{n-m}\binom{n-m}{m}(2x)^{n-2m} \qquad (1.8)$$

$$\left.\right\} \; n > 0$$

Binet forms are, respectively,

$$P_n(x) = (\alpha^n - \beta^n)/(\alpha - \beta) \qquad (1.9)$$

and

$$Q_n(x) = \alpha^n + \beta^n, \qquad (1.10)$$

where

$$\begin{cases} \alpha = x + \sqrt{x^2 + 1} \\[2mm] \beta = x - \sqrt{x^2 + 1} \end{cases} \qquad (1.11)$$

are the roots of

$$t^2 - 2xt - 1 = 0, \qquad (1.12)$$

so that

$$\alpha + \beta = 2x, \quad \alpha\beta = -1, \quad \alpha - \beta = 2\sqrt{x^2 + 1}. \qquad (1.13)$$

Expressions for $P_n(x)$ and $Q_n(x)$ for $n = 0, 1, \ldots, 5$ are tabulated below for ready reference in Table 1.

Table 1. Pell and Pell-Lucas Polynomials

$n = 0$	1	2	3	4	5
$P_n(x)$	0	1	$2x$	$4x^2 + 1$	$8x^3 + 4x$	$16x^4 + 12x^2 + 1$
$Q_n(x)$	2	$2x$	$4x^2 + 2$	$8x^3 + 6x$	$16x^4 + 16x^2 + 2$	$32x^5 + 40x^3 + 10x$

Properties of $P_n(x)$ and $Q_n(n)$ with values of n extended to negative integers where necessary, are set out in [13], [20], and [21]. Analytic aspects of $P_n(x)$ are treated in [2], in a different notation, and extended in [30]. Emphasis throughout this paper will be placed on $P_n(x)$ rather than on the more complicated $Q_n(x)$, though the properties of

both are intertwined, as we have come to expect from their
Fibonacci and Lucas counterparts.

Information about convolutions for Fibonacci polynomials can be found in, say, [7] and [28]. Undoubtedly, the
approach in these sources could have been followed, but we
have elected to pursue our investigation independently of
them.

2. Some Properties of Pell and Pell-Lucas Polynomials

Among the properties of $P_n(x)$ and $Q_n(x)$ that we shall subsequently require are the following, listed roughly in order
of use:

$$\frac{dQ_n(x)}{dx} = 2nP_n(x); \tag{2.1}$$

$$\sum_{n=1}^{\infty} nP_{n+1}(x)y^{n-1} = (2x + 2y)(1 - 2xy - y^2)^{-2}; \tag{2.2}$$

$$\sum_{n=1}^{\infty} nQ_{n+1}(x)y^{n-1} = (4x^2 + 2 + 4xy + 2y^2) \times$$
$$\times (1 - 2xy - y^2)^{-2}; \tag{2.3}$$

$$P_{n+r}(x) + P_{n-r}(x) = \begin{cases} P_n(x)Q_r(x) & r \text{ even,} \\ Q_n(x)P_r(x) & r \text{ odd;} \end{cases} \tag{2.4}$$

$$P_{n+r}(x) - P_{n-r}(x) = \begin{cases} Q_n(x)P_r(x) & r \text{ even,} \\ P_n(x)Q_r(x) & r \text{ odd;} \end{cases} \tag{2.5}$$

$$Q_{n+r}(x) + Q_{n-r}(x) = \begin{cases} Q_n(x)Q_r(x) & r \text{ even,} \\ 4(x^2 + 1)P_n(x)P_r(x) & r \text{ odd;} \end{cases} \tag{2.6}$$

$$Q_{n+r}(x) - Q_{n-r}(x) = \begin{cases} 4(x^2 + 1)P_n(x)P_r(x) & r \text{ even,} \\ Q_n(x)Q_r(x) & r \text{ odd;} \end{cases} \tag{2.7}$$

$$P_{n+1}(x)P_{n-1}(x) - P_n^2(x) = (-1)^n; \tag{2.8}$$

$$Q_{n+1}(x)Q_{n-1}(x) - Q_n^2(x) = (-1)^{n-1}4(x^2 + 1); \tag{2.9}$$

Properties (2.8) and (2.9) are *Simson formulas*.

$$P_{n-1}(x)Q_{n+2}(x) - P_{n+1}(x)Q_n(x) = 4x^2(-1)^n; \qquad (2.10)$$

$$P_{n-1}(x)Q_{n+2}(x) + P_{n+1}(x)Q_n(x) - 2P_n(x)Q_{n+1}(x)$$
$$= (-1)^n 4(x^2 + 1); \qquad (2.11)$$

$$Q_n(x) = 2xP_n(x) + 2P_{n-1}(x)$$
$$= P_{n+1}(x) + P_{n-1}(x) \qquad \text{by (1.3)} \qquad (2.12)$$
[case (2.4) when $r = 1$].

3. Gegenbauer, Chebyshev, and Generalized Humbert Polynomials

Gegenbauer and Chebyshev Polynomials. The *Gegenbauer polynomials* $C_n^\lambda(x)$ of degree n and order λ may be defined by the generating function

$$(1 - 2xy + y^2)^{-\lambda} = \sum_{n=0}^{\infty} C_n^\lambda(x)y^n \qquad (\lambda > 0, \ |y| < 1). \qquad (3.1)$$

Chebyshev polynomials of the second kind $U_n(x)$ arise as a special case when $\lambda = 1$, i.e.,

$$U_n(x) = C_n^1(x). \qquad (3.2)$$

In [13] it is shown that

$$\begin{cases} P_n(x) = (-i)^{n-1}C_{n-1}^1(ix) \\ \\ \quad = (-i)^{n-1}U_{n-1}(ix) \qquad \text{by (3.2).} \end{cases} \qquad (3.3)$$

Putting $y = P_{n+1}(x)$, we find that the differential equation satisfied by Pell polynomials is

$$(1 + x^2)\frac{d^2y}{dx^2} + 3x\frac{dy}{dx} - n(n + 2)y = 0. \qquad (3.4)$$

This follows immediately from the known ([4], [19], [23]) differential equation satisfied by Gegenbauer and Chebyshev polynomials, when we invoke the use of (3.3). See also Section 6 below, where convolutions are employed to establish (3.4).

Explicit summation formulas for $C_n^\lambda(x)$ and $U_n(x)$ are

$$C_n^\lambda(x) = \sum_{j=0}^{[n/2]} (-1)^j \binom{\lambda + n - 1 - j}{\lambda - 1}\binom{n-j}{j}(2x)^{n-2j}$$

$$(\lambda > 0), \quad (3.5)$$

$$C_n^0(x) = \sum_{j=0}^{[n/2]} \frac{(-1)^j}{n-j}\binom{n-j}{j}(2x)^{n-2j} \quad C_0^0(x) = 1, \quad (3.6)$$

and

$$U_n(x) = \sum_{j=0}^{[n/2]} (-1)^j \binom{n-j}{j}(2x)^{n-2j}, \quad (3.7)$$

where we note that in (3.5) the binomial factor

$$\binom{\lambda + n - 1 - j}{\lambda - 1} = \frac{\Gamma(n + \lambda - j)}{\Gamma(\lambda)\Gamma(n + 1 - j)}$$

in terms of Gamma functions.

Chebyshev polynomials of the first kind $T_n(x)$ are given explicitly by

$$T_n(x) = \frac{1}{2}\sum_{j=0}^{[n/2]} (-1)^j \frac{n}{n-j}\binom{n-j}{j}(2x)^{n-2j}. \quad (3.8)$$

It was shown in [13] that

$$Q_n(x) = 2(-i)^n T_n(ix) = (-i)^n C_n^0(ix), \quad (3.9)$$

since

$$C_n^0(x) = \frac{2}{n} T_n(x) \quad \text{from (3.6) and (3.8).}$$

Known results for $U_n(x)$ and $T_n(x)$, when x is replaced by ix, produce results for $P_n(x)$ and $Q_n(x)$. For instance,

$$\frac{dT_n(x)}{dx} = nU_{n-1}(x)$$

leads to

$$\frac{dQ_n(x)}{dx} = 2nP_n(x) \quad \text{(see [13]).}$$

Additionally,

$$\begin{cases} T_{n+r}(x) + T_{n-r}(x) = 2T_n(x)T_r(x) \quad \text{and} \\ T_{n+r}(x) - T_{n-r}(x) = 2(x^2 - 1)U_{n-1}(x)U_{r-1}(x) \end{cases}$$

yield

$$\begin{cases} Q_{n+r}(x) \mp Q_{n-r}(x) = Q_n(x)Q_r(x) & \begin{cases} r \text{ odd} \\ r \text{ even} \end{cases} \text{ and} \\ Q_{n+r}(x) \pm Q_{n-r}(x) = 4(x^2+1)P_n(x)P_r(x) & \begin{cases} r \text{ odd} \\ r \text{ even} \end{cases} \end{cases}$$

respectively, as in (2.6) and (2.7).

Hypergeometric functions $_2F_1(\alpha, \beta; \gamma, x)$ are defined in [27] by

$$_2F_1(\alpha, \beta; \gamma; x) = \sum_{r=0}^{\infty} \frac{(\alpha)_r (\beta)_r}{(\gamma)_r r!} x^r$$

$$= {}_2F_1(\beta, \alpha; \gamma; x) \qquad (3.10)$$

in which

$$(\alpha)_r = \alpha(\alpha + 1)(\alpha + 2) \cdots (\alpha + r - 1)$$

$$= \frac{\Gamma(\alpha + r)}{\Gamma(\alpha)} \qquad (3.11)$$

and

$$_2F_1(\alpha, \beta; \gamma; 0) = 1. \qquad (3.12)$$

Hypergeometric representations of $C_n^\lambda(x)$, $U_n(x)$, and $T_n(x)$ are:

$$C_n^\lambda(x) = \frac{\Gamma(n + 2\lambda)}{n!\Gamma(2\lambda)} {}_2F_1\left(-n, n + 2\lambda; \lambda + \frac{1}{2}; \frac{1-x}{2}\right); \qquad (3.13)$$

$$U_n(x) = (n+1){}_2F_1\left(-n, n + 2; \frac{3}{2}; \frac{1-x}{2}\right); \qquad (3.14)$$

$$T_n(x) = {}_2F_1\left(-n, n; \frac{1}{2}; \frac{1-x}{2}\right). \qquad (3.15)$$

Let us note in passing that when $\lambda = 1/2$ the *Legendre polynomials* $C_n^{1/2}(x)$ occur. The Gegenbauer (or *ultraspherical*) polynomials are themselves a special case of the *Jacobi polynomials* ([4], [19], [23]).

Other information about Gegenbauer polynomials can be found in [5], [12], and [14]. Of the many texts on Chebyshev polynomials, [24] provides a useful background for our purposes.

Furthermore, we note that in (3.11) the rising factorials $(\alpha)_r$, considered as polynomials in α, are the generating functions for the *Stirling numbers of the first kind* (see [4]).

Combinatorial identities relating to (3.11) which are needed for subsequent reference are herewith appended:

$$\binom{-k-1}{n} = (-1)^n\binom{k+n}{k} = (-1)^n\,\frac{(k+1)_n}{n!}; \qquad (3.16)$$

$$\binom{-k-1}{k} = (-1)^k\binom{2k}{k} \qquad [n = k \text{ in } (3.16)] \qquad (3.17)$$

$$K_k = \frac{1}{k+1}\binom{2k}{k}; \qquad (3.18)$$

where K_k is the kth *Catalan number*. Thus, by (3.11) and (3.18),

$$(k+1)_k = (k+1)!K_k. \qquad (3.19)$$

The binomial coefficient $\binom{n+k}{k}$ occurring in (3.16) is the number of partitions of n into $k+1$ nonnegative summands, a result due to Catalan.

Generalized Humbert Polynomials. Concluding this section, we introduce the *generalized Humbert polynomial* $P_n(m, x, y, p, C)$ defined in [6] by

$$(C - mxt + yt^m)^p = \sum_{n=0}^{\infty} P_n(m, x, y, p, C)t^n, \qquad (3.20)$$

where $m \geqslant 1$ is an integer and the other parameters are unrestricted in general.

Besides Humbert's original polynomials, some other polynomials included in (3.20) as special cases are those of Chebyshev, Gegenbauer, Legendre, and Pincherle. Specific details of these can be found in [6] and [12].

More particularly, we note that Pell polynomials are included in the generalized Humbert system when $p = -1$; namely, by (1.5) and (3.20),

$$P_{n+1}(x) = P_n(2, x, -1, -1, 1) \qquad (n = 0, 1, 2, \ldots), \qquad (3.21)$$

with

$$P_1(x) = P_0(2, x, -1, -1, 1) = 1,$$
$$P_0(x) = 0 \quad \text{as in } (1.3). \qquad (3.22)$$

By using (when $k = 0$) the inversion formula established by Gould [6] for generalized Humbert polynomials [see (9.3)], we have *mutatis mutandis* the inversion formula for Pell polynomials,

$$(2x)^n = \sum_{j=0}^{[n/2]} (-1)^j \binom{n}{j} \frac{n + 1 - 2j}{n + 1 - j} \, P_{n+1-2j}(x), \quad (3.23)$$

a result given earlier in [2] in a slightly varied notation.

Subsequently, we make minor reference to the *Pincherle polynomials* $\mathscr{P}_n(x)$ (see [15], [22]) given by

$$\mathscr{P}_n(x) = P_n\left(3, \, x, \, 1, \, -\frac{1}{2}, \, 1\right). \quad (3.24)$$

Furthermore, the ubiquitous *Fibonacci numbers* F_n, occurring when $x = 1/2$, are

$$F_{n+1} = P_n\left(2, \, \frac{1}{2}, \, -1, \, -1, \, 1\right) = \sum_{k=0}^{[n/2]} \binom{n-k}{k}. \quad (3.25)$$

Unlike the Chebyshev and Gegenbauer polynomials, the Pell polynomials $P_n(x)$ are not orthogonal on any interval of the real axis (see [2]). For other cognate remarks, refer to [6] in the case of generalized Humbert polynomials when $m = 2$ (Section 9), and [26] in the case of a generalized Fibonacci polynomial analogous to those of Bernoulli, Euler, and Hermite. Details of orthogonal polynomials can be found, for example, in [23] and [29], while the appropriate weight functions are also listed in [4] for the classical orthogonal polynomials of Legendre, Gegenbauer, Chebyshev, Jacobi, Hermite, and (in the generalized case) Laguerre.

Interestingly, one of the motivations stated in [6] for generalizing Legendre polynomials lies in the desire to study generalized potential problems.

PART II: CONVOLUTIONS

4. First Convolution Sequence for Pell Polynomials

The *first convolution sequence for Pell polynomials*

$$\{P_n^{(1)}(x)\}, \quad n = 1, 2, 3, \ldots,$$

is defined by

$$P^{(1)}(x) = \sum_{j=1}^{n} P_j(x)P_{n+1-j}(x), \quad P_0^{(1)}(x) = 0, \qquad (4.1)$$

with generating function, from (1.5),

$$(1 - 2xy - y^2)^{-2} = \sum_{n=0}^{\infty} P_{n+1}^{(1)}(x)y^n. \qquad (4.2)$$

From (4.1), we derive the basic recursion-type result:

$$P_{n+1}^{(1)}(x) - 2xP_n^{(1)}(x) - P_{n-1}^{(1)}(x) = P_{n+1}(x). \qquad (4.3)$$

Partial differentiation of both sides of (1.5) with respect to y followed by the use of (4.2) on the left-hand side together with the equating of the coefficients of y^{n-1} yields

$$nP_{n+1}(x) = 2xP_n^{(1)}(x) + 2P_{n-1}^{(1)}(x). \qquad (4.4)$$

Combining (4.3) and (4.4), we obtain the recursive relation

$$nP_{n+1}^{(1)}(x) = 2x(n + 1)P_n^{(1)}(x) + (n + 2)P_{n-1}^{(1)}(x), \qquad (4.5)$$

whence

$$(n + 1)P_{n+1}(x) = P_{n+1}^{(1)}(x) + P_{n-1}^{(1)}(x). \qquad (4.6)$$

Next, from (4.2),

$$\sum_{n=1}^{\infty} P_n^{(1)}(x)y^{n-1}$$

$$= \frac{(4x^2 + 2) + 4xy + 2y^2 + 2(1 - 2xy - y^2)}{4(x^2 + 1)(1 - 2xy - y^2)^2}$$

$$= \frac{1}{4(x^2 + 1)} \sum_{n=1}^{\infty} \{nQ_{n+1}(x) + 2P_n(x)\}y^{n-1}$$

by (1.5) and (2.3).

Comparing coefficients of y^{n-1}, we deduce

$$P_n^{(1)}(x) = \{nQ_{n+1}(x) + 2P_n(x)\}/4(x^2 + 1). \qquad (4.7)$$

Employing (1.9) and (1.10) in conjunction with (4.7) gives us a Binet-type form for $P_n^{(1)}(x)$. Equations (6.1) through (6.4) of [13] also allow us to express $P_n^{(1)}(x)$ in (4.7) in terms of hyperbolic functions, if we put $x = \sinh t$.

Using (4.3) as a difference equation, we may obtain the summation formula:

$$\sum_{r=1}^{n} P_r^{(1)}(x) = \left\{ P_{n+2}^{(1)}(x) + P_{n+1}^{(1)}(x) - P_n^{(1)}(x) - P_{n-1}^{(1)}(x) - \right.$$
$$\left. - 2P_{n+2}(x) - 2P_{n+1}(x) + 1 \right\}/4x^2. \quad (4.8)$$

An explicit formula for $P_n^{(1)}(x)$ is

$$P_n^{(1)}(x) = \sum_{r=0}^{[(n-1)/2]} \binom{n-r}{1}\binom{n-r-1}{r}(2x)^{n-1-2r}. \quad (4.9)$$

For example, when $n = 6$, we have $P_6^{(1)}(x) = 192x^5 + 160x^3 + 24x$. See Table 2 for other values of n.

To prove (4.9), we appeal to the method of mathematical induction, using (4.3) and Pascal's formula. Cases n even, n odd need to be considered separately.

Differentiating in (1.7) and applying (4.9) establishes

$$P_{n+1}'(x) = 2P_n^{(1)}(x) \qquad\qquad\qquad (4.10)$$

whence, by (2.1),

$$Q_n''(x) = 4nP_{n-1}^{(1)}(x), \qquad\qquad\qquad (4.11)$$

in which the primes denote differentiation with respect to x.

Equation (4.10) may be neatly obtained from (1.5) by differentiating partially with respect to x and then equating coefficients of y^r on both sides. Differentiation w.r.t. x in (1.3) followed by use of (4.10) leads us back to (4.3).

If in (1.9), with $n + 1$ substituted for n, we differentiate with respect to x, we arrive back at (4.7) after simplification using (4.10), (1.9), and (1.10).

Replacing x by ix ($i = \sqrt{-1}$) in (3.5) and using (4.9), we obtain a complex expression for the first convolution of Pell polynomials in terms of Gegenbauer polynomials, namely,

$$C_n^2(ix) = i^n P_{n+1}^{(1)}(x), \qquad n \geqslant 0. \qquad\qquad (4.12)$$

Using the hypergeometric representation for the Gegenbauer polynomial (3.13) with x replaced by $-i$, we have:

$$P_n^{(1)}(-i) = \frac{i^{-n+1}(n+2)!}{3!(n-1)!}.$$ (4.13)

From the hyperbolic expressions for the Pell polynomials given in equations (6.1) and (6.2) of [13], we readily obtain, by differentiation and application of (4.10), hyperbolic expressions for $P_n^{(1)}(x)$, where $x = \sinh t$ in the two cases—n even, n odd.

Invoking (4.7) and (2.8)-(2.11), we establish the analogue of *Simson's formula*, namely,

$$P_{n+1}^{(1)}(x)P_{n-1}^{(1)}(x) - [P_n^{(1)}(x)]^2$$

$$= \frac{(-1)^n\{n(n+2)4(x^2+1) + 4x^2\} - Q_{n+1}^2(x)}{[4(x^2+1)]^2}.$$ (4.14)

Other properties of $P_n^{(1)}(x)$ are given in [13].

5. First Convolution Sequence for Pell-Lucas Polynomials

Turning now to the polynomials $Q_n(x)$, we define the *first convolution sequence for Pell-Lucas polynomials*

$$\{Q_n^{(1)}(x)\}, \quad n = 1, 2, 3, \ldots,$$

by

$$Q_n^{(1)}(x) = \sum_{j=1}^{n} Q_j(x)Q_{n+1-j}(x), \qquad Q_0^{(1)}(x) = 0,$$ (5.1)

with generating function, from (1.5),

$$\left(\frac{2x+2y}{1-2xy-y^2}\right)^2 = \sum_{n=0}^{\infty} Q_{n+1}^{(1)}(x)y^n.$$ (5.2)

It follows immediately from (4.2) and (5.2) that

$$Q_n^{(1)}(x) = 4x^2 P_n^{(1)}(x) + 8x P_{n-1}^{(1)}(x) + 4P_{n-2}^{(1)}(x).$$ (5.3)

Application of (5.1) in the manner of (4.3) yields

$$Q_{n+1}^{(1)}(x) = 2x Q_n^{(1)}(x) + Q_{n-1}^{(1)}(x) + 4(x^2+1)P_{n+1}(x).$$ (5.4)

Now (5.2) with (2.2) leads to

$$\sum_{n=1}^{\infty} Q_n^{(1)}(x) y^{n-1} = (2x + 2y) \sum_{n=1}^{\infty} n P_{n+1}(x) y^{n-1}$$

$$= \sum_{n=1}^{\infty} \{2xn P_{n+1}(x) + 2(n-1)P(x)\} y^{n-1}$$

$$= \sum_{n=1}^{\infty} \{n Q_{n+1}(x) - 2P_n(x)\} y^{n-1},$$

by (2.12), whence

$$Q_n^{(1)}(x) = n Q_{n+1}(x) - 2P_n(x), \qquad (5.5)$$

which allows us to derive a Binet-type representation for $Q_n^{(1)}(x)$ by (1.9) and (1.10) [cf. (4.7)].

Equations (1.7), (1.8), and (5.5) produce, on calculation,

$$Q_n^{(1)}(x) = \sum_{r=0}^{[(n+1)/2]} \frac{1}{n+1-r} \binom{n+1-r}{r} \times$$

$$\times \{n(n+1) - 2r\}(2x)^{n+1-2r}. \qquad (5.6)$$

Combining (4.7) and (5.5), we find

$$4(x^2 + 1)P_n^{(1)}(x) + Q_n^{(1)}(x) = 2n Q_{n+1}(x) \qquad (5.7)$$

and

$$4(x^2 + 1)P_n^{(1)}(x) - Q_n^{(1)}(x) = 4P_n(x). \qquad (5.8)$$

Moreover, (4.7) also has the consequence (4.6), namely, that

$$P_{n+1}^{(1)}(x) + P_{n-1}^{(1)}(x) = (n+1)P_{n+1}(x)$$

after appeal successively to (2.6), (1.3), (2.4), and (2.6) again.

More generally, with recourse to (2.4)-(2.7), (4.7), and (5.5), we may establish the following formulas:

$$P_{n+m}^{(1)}(x) + (-1)^m P_{n-m}^{(1)}(x)$$

$$= P_n^{(1)}(x) Q_m(x) + m P_{n+1}(x) P_m(x); \qquad (5.9)$$

$$P_{n+m}^{(1)}(x) - (-1)^m P_{n-m}^{(1)}(x)$$

$$= \{4n(x^2 + 1)P_{n+1}(x)P_m(x) + mQ_{n+1}(x)Q_m(x) +$$

$$+ 2Q_n(x)P_m(x)\}/4(x^2 + 1); \qquad (5.10)$$

$$Q_{n+m}^{(1)}(x) + (-1)^m Q_{n-m}^{(1)}(x)$$

$$= Q_n^{(1)}(x)Q_m(x) + 4m(x^2 + 1)P_{n+1}(x)P_m(x); \qquad (5.11)$$

$$Q_{n+m}^{(1)}(x) - (-1)^m Q_{n-m}^{(1)}(x)$$

$$= 4n(x^2 + 1)P_{n+1}(x)P_m(x) + mQ_{n+1}(x)Q_m(x)$$

$$- 2Q_n(x)P_m(x). \qquad (5.12)$$

By means of (5.3), (4.6), (1.3), and (2.12) used successively [or by (5.12) with $m = 1$, and (1.2) and (2.2)], we
may demonstrate that

$$Q_{n+1}^{(1)}(x) + Q_{n-1}^{(1)} = 4(n + 1)(x^2 + 1)P_{n+1}(x) - 4Q_n(x).$$

$$(5.13)$$

On the authority of (5.5) and (2.8)-(2.11) we establish, as in (4.14), the analogue of the *Simson formula*

$$Q_{n+1}^{(1)}(x)Q_{n-1}^{(1)}(x) - [Q_n^{(1)}(x)]^2$$

$$= -Q_{n+1}^2(x) + (-1)^n n(n - 2)4(x^2 + 1) - (-1)^n 12x^2.$$

$$(5.14)$$

No effort is required to show that

$$P_n^{(1)}(x)Q_n^{(1)}(x) = \frac{n^2 Q_{n+1}^2(x) - 4P_n^2(x)}{4(x^2 + 1)}, \qquad (5.15)$$

on the basis of (4.7) and (5.5).
Obtaining new relationships among the Pell and Pell-
Lucas polynomials is an almost endless process. We conclude
this section with the rather nice summation result

$$(n + 1)P_{n+2}^{(1)}(x) = 2 \sum_{r=0}^{n} P_{r+1}^{(1)}(x)Q_{n+1-r}(x), \qquad (5.16)$$

which may be derived from (4.2) by differentiating w.r.t. y and then reassembling the coefficients of y^n.

For reference purposes, we append in Table 2 the expressions for $n = 1, 2, \ldots, 5$ for $P_n^{(1)}(x)$ and $Q_n^{(1)}(x)$, and for $P_n^{(2)}(x)$ and $Q_n^{(2)}(x)$, which are to be defined in the next section.

Table 2. Convolutions for $P^{(k)}(x)$, $Q^{(k)}(x)$, $k = 1, 2$;
$n = 1, 2, 3, 4, 5$

	$n = 1$	2	3
$P^{(1)}(x)$	1	$4x$	$12x^2 + 2$
$Q^{(1)}(x)$	$4x^2$	$16x^3 + 8x$	$48x^4 + 40x^2 + 4$
$P^{(2)}(x)$	1	$6x$	$24x^2 + 3$
$Q^{(2)}(x)$	$8x^3$	$48x^4 + 24x^2$	$192x^5 + 168x^3 + 24x$
	$n = 1$	4	5
$P^{(1)}(x)$	1	$32x^3 + 12x$	$80x^4 + 48x^2 + 3$
$Q^{(1)}(x)$	$4x^2$	$128x^5 + 144x^3 + 32x$	$320x^6 + 488x^4 + 156x^2 + 8$
$P^{(2)}(x)$	1	$80x^3 + 24x$	$240x^4 + 120x^2 + 6$
$Q^{(2)}(x)$	$8x^3$	$640x^6 + 768x^4 + 216x^2 + 8$	$1920x^7 + 2880x^5 + 1200x^3 + 120x$

6. Second Convolution Sequences
$$\left\{ P^{(2)}(x) \right\}, \ \left\{ Q^{(2)}(x) \right\}$$

Define the *second convolution sequence for Pell polynomials*

$$\{P_n^{(2)}(x)\}, \quad n = 1, 2, 3, \ldots,$$

by

$$P_n^{(2)}(x) = \sum_{j=1}^{n} P_j(x) P_{n+1-j}^{(1)}(x), \qquad P_0^{(2)}(x) = 0. \qquad (6.1)$$

A generating function for the sequence is

$$(1 - 2xy - y^2)^{-3} = \sum_{n=1}^{\infty} P_{n+1}^{(2)}(x) y^n. \qquad (6.2)$$

With a development similar to that for $P_n^{(1)}(x)$, we may establish:

$$P_n^{(2)}(x) = 2xP_{n-1}^{(2)}(x) + P_{n-2}^{(2)}(x) + P_n^{(1)}(x); \qquad (6.3)$$

$$nP_{n+1}^{(1)}(x) = 4xP_n^{(2)}(x) + 4P_{n-1}^{(2)}(x); \qquad (6.4)$$

$$nP_{n+1}^{(2)}(x) = 2x(n+2)P_n^{(2)}(x) + (n+4)P_{n-1}^{(2)}(x); \quad (6.5)$$

$$P_n^{(2)}(x) = \sum_{r=0}^{[(n-1)/2]} \binom{n+1-r}{2}\binom{n-1-r}{r}(2x)^{n-1-2r}; \qquad (6.6)$$

$$P_{n+2}''(x) = 2^2 \cdot 2!P_n^{(2)}(x). \qquad (6.7)$$

By (6.4) and (6.7) it follows that

$$x\{P_{n+1}^{(1)}(x)\}' = nP_{n+1}^{(1)}(x) - \{P_n^{(1)}(x)\}', \qquad (6.8)$$

i.e., a formula involving first convolutions of $P_n(x)$.

Furthermore, reverting to (3.5) and then comparing with (6.6) gives us the formula for third-order Gegenbauer polynomials in the complex form

$$C_n^3(ix) = i^n P_{n+1}(x), \qquad (6.9)$$

whence, after substitution in the appropriate hypergeometric function,

$$P_n^{(2)}(-i) = \frac{i^{-n+1}(n+4)!}{5!(n-1)!} \qquad (6.10)$$

Applications of (4.10), (6.3), (4.4), (6.4), and (6.7) combine to provide us with another verification of a solution of the differential equation (3.4).

Other properties of $P_n^{(2)}(x)$ might be investigated, but we do not pursue this course of action here.

Expressions for $P_n^{(2)}(x)$ when $n = 1, 2, \ldots, 5$ are given in Table 2 above.

Next, we define *the second convolution sequence for Pell-Lucas polynomials*

$$\{Q_n^{(2)}(x)\}, \quad n = 1, 2, 3, \ldots,$$

by

$$Q_n^{(2)}(x) = \sum_{j=1}^{n} Q_j(x)Q_{n+1-j}^{(1)}(x), \qquad Q_0^{(2)}(x) = 0, \quad (6.11)$$

with generating function

$$\left(\frac{2x + 2y}{1 - 2xy - y^2}\right)^3 = \sum_{n=0}^{\infty} Q_{n+1}^{(2)}(x) y^n. \qquad (6.12)$$

Table 2 gives expressions for $Q_1^{(2)}(x)$, ..., $Q_5^{(2)}(x)$. From (6.2) and (6.12), it follows that

$$Q_n^{(2)}(x) = 8x^3 P_n^{(2)}(x) + 24x^2 P_{n-1}^{(2)}(x) + 24x P_{n-2}^{(2)}(x) +$$
$$+ 8P_{n-3}^{(2)}(x)$$

$$= \frac{n}{2} Q_{n+1}^{(1)}(x) - 2n P_{n+1}(x) \qquad \text{by (4.4), (5.3),}$$
$$\text{and (6.4)}$$

$$= \frac{n(n+1)}{2} Q_{n+2}(x) - 3n P_{n+1}(x) \qquad \text{by (5.5).}$$
$$(6.13)$$

which corresponds to (5.5) for $Q_n^{(1)}(x)$.

Using (1.9), (1.10), and (6.13), we can obtain a Binet form for $Q_n^{(2)}(x)$.

Desirably, an expression for $P_n^{(2)}(x)$ in terms of $P_n(x)$ and $Q_n(x)$ to match (6.13) is needed. Unfortunately, the existence of such an expression has not yet been forthcoming. [Cf. the formula (4.7) for $P_n^{(1)}(x)$.]

7. k^{th} Convolution Sequence $\left\{P_n^{(k)}(x)\right\}$

Generalizing the previous ideas, we define the kth *convolution sequence for Pell polynomials*

$$\{P_n^{(k)}(x)\}, \quad n = 1, 2, 3, \ldots,$$

by the equivalent expressions

$$P_n^{(k)}(x) = \begin{cases} \sum_{i=1}^{n} P_i(x) P_{n+1-i}^{(k-1)}(x) & k \geqslant 1 \\[2ex] \sum_{i=1}^{n} P_i^{(1)}(x) P_{n+1-i}^{(k-2)}(x) & \\[2ex] \cdots\cdots\cdots\cdots\cdots & \\[2ex] \sum_{i=1}^{n} P_i^{(m)}(x) P_{n+1-i}^{(k-1-m)}(x) & 0 \leqslant m \leqslant k-1 \end{cases} \qquad \begin{cases} P_n^{(0)}(x) = P_n(x) \\[2ex] P_0^{(k)}(x) = 0 \end{cases} \qquad (7.1)$$

for which the generating function is

$$(1 - 2xy - y^2)^{-(k+1)} = \sum_{n=0}^{\infty} P_{n+1}^{(k)}(x) y^n. \qquad (7.2)$$

For example, from (7.1),

$$P_2^{(3)}(x) = \begin{Bmatrix} P_1(x)P_2^{(2)}(x) + P_2(x)P_1^{(1)}(x) = 6x + 2x \\ 2P_1^{(1)}(x)P_2^{(1)}(x) = 2.4x \end{Bmatrix} = 8x,$$

as may be readily verified from (7.2).
The first few values of $P_n^{(k)}(x)$ are

$$\begin{cases} P_1^{(k)}(x) = 1, \ P_2^{(k)}(x) = (k + 1)2x, \\ P_3^{(k)}(x) = (k + 1)(k + 2)2x^2 + k + 1 \\ \\ P_4^{(k)}(x) = \dfrac{(k + 1)(k + 2)(k + 3)}{3!} 8x^3 + \\ \qquad + (k + 1)(k + 2)2x, \ \ldots . \end{cases}$$

Interested readers may wish to calculate some of the expressions for $P_n^{(3)}(x)$ for small values of n, as these are sometimes convenient in illustrative computations.
Induction with (4.1) leads to the recursion formula

$$P_{n+1}^{(k)}(x) = 2xP_n^{(k)}(x) + P_{n-1}^{(k)}(x) + P_{n+1}^{(k-1)}(x). \qquad (7.3)$$

Differentiating both sides of (7.2) with respect to y gives us

$$nP_{n+1}^{(k)}(x) = 2(k + 1)\left\{xP_n^{(k+1)}(x) + P_{n-1}^{(k+1)}(x)\right\}. \qquad (7.4)$$

When $k = 0$, $k = 1$ we obtain (4.4) and (6.4), respectively.
Further, if we reassemble the coefficients of y^{n-1} on the differentiated left–hand side of (7.2), we establish that

$$nP_{n+1}^{(k)}(x) = (k + 1)\sum_{i=1}^{n} Q_i(x)P_{n+1-i}^{(k)}(x). \qquad (7.5)$$

Combining (7.3) and (7.4) produces

$$nP_{n+1}^{(k)}(x) = 2x(n + k)P_n^{(k)}(x) + (n + 2k)P^{(k)}(x). \qquad (7.6)$$

Notice that as $n \to \infty$, (7.6) approximates to the basic Pell-type recurrence (1.3), assuming k remains small in relation to n.

An explicit formula for the kth Pell convolution polynomial may be established by induction to be

$$P_n^{(k)}(x) = \sum_{r=0}^{[(n-1)/2]} \binom{k+n-1-r}{k}\binom{n-1-r}{r}(2x)^{n-2r-1}.$$

(7.7)

Putting $k = 1$ gives (4.9), while $k = 2$ gives (6.6).

Differentiation of equation (7.7) with respect to x yields

$$\frac{d}{dx} P_n^{(k)}(x) = 2(k+1)P_{n-1}^{(k+1)}(x)$$

(7.8)

after some algebraic manipulation.

Replacing r by $n + k$ in (1.5) and then differentiating k times with respect to x, we deduce that

$$\frac{d^k}{dx^k}\{P_{n+k}(x)\} = 2^k \cdot k! P_n^{(k)}(x).$$

(7.9)

Gegenbauer polynomials are, by (3.1) and (7.2), related to the kth Pell convolution polynomial by

$$C_n^k(ix) = i^n P_{n+1}^{(k-1)}(x)$$

(7.10)

whence $P_{n+1}^{(k)}(x)$ may be expressed hypergeometrically from which, as a special case,

$$P_n^{(k)}(-i) = \frac{i^{-n+1}(n+2k)!}{(2k+1)!(n-1)!}$$

Taken together, (3.5) and (7.10) supply us with a verification for the validity of (7.7).

Moreover, (7.9) may also be obtained from (3.3), (7.10), and the known result (see [4] and [19]) connecting Chebyshev and Gegenbauer polynomials,

$$\frac{d^m}{dx^m} U_n(x) = 2^m m! C_{n-m}^{m+1}(x), \qquad n > m.$$

With the aid of (3.9) and (7.10) in conjunction with the known result (see [4] and [19]),

$$\frac{d^m}{dx^m} T_n(x) = 2^{m-1}\Gamma(m)nC_{n-m}^m(x), \qquad n > m,$$

we can demonstrate that

$$\frac{d^k}{dx^k} Q_n(x) = 2^k n\Gamma(k)P_{n-k+1}^{(k-1)}(x), \qquad n+1 > k, \qquad (7.12)$$

of which (4.11) is the special case occurring when $k = 2$.

Further information about $P_n^{(k)}(x)$ can be found in [13], among which are some complicated expressions for $P_n^{(k)}(x)$.

Among other results of interest in this source are the formulas

$$(n)_k P_{n+k}(x) = k! \sum_{i=0}^{k+1} \binom{k+1}{i} P_{k+1-i}(x)P_{n-i}^{(k)}(x) \qquad (7.13)$$

and

$$(n)_k Q_{n+k}(x) = k! \sum_{i=0}^{k+1} \binom{k+1}{i} Q_{k+1-i}(x)P_{n-i}^{(k)}(x), \qquad (7.14)$$

where $(n)_k$ is defined in (3.11).

These results are obtained by differentiating n times both sides of equations (1.5) and (1.6), respectively. Note the slight similarity between the forms of (7.1) and (7.13).

Although we have concentrated in this paper on $P_n^{(k)}(x)$ where $n \geqslant 0$ (integer), the theory would be incomplete without brief mention of negative subscripts. Appealing to (7.3) and (7.6), we may establish by induction that

$$P_{-n}^{(k)}(x) = (-1)^{n-k-1}P_{n-2k}^{(k)}(x). \qquad (7.15)$$

8. Use of Differential Equations Involving Gegenbauer Polynomials

Many of the results in the preceding section may be obtained from known differential equations involving the Gegenbauer polynomials, with complex variable ix, used in conjunction with (7.10).

Details are suppressed to conserve space.

We observe in passing that the use of (7.10), (7.8), and then (7.10) again leads to

$$\frac{d}{dx} C_n^k(ix) = i \cdot 2kC_{n-1}^{k+1}(ix),$$

which is the complex version of a known result ([14],[19]).

Relationships not previously stated among Pell polynomials may be established by using corresponding differential equations involving Gegenbauer polynomials. For instance, the differential equation (see [19])

$$\frac{d}{dx}[C_{n+1}^k(x) - C_{n-1}^k(x)] = 2(n + k)C_n^k(x) \qquad (8.1)$$

or, as given equivalently in [4] in integral form,

$$C_{n+1}^k(x) - C_{n-1}^k(x) = 2(n + k)\int C_n^k(x)dx \qquad (8.2)$$

leads to

$$(n + k)P_{n+1}^{(k-1)}(x) = k\{P_{n+1}^{(k)}(x) + P_{n-1}^{(k)}(x)\}, \qquad (8.3)$$

a result [also obtainable from (7.3) and (7.4)] which we find attractive. Equation (4.6) is a special case of this when $k = 1$. The special case for $k = 2$ may also be obtained from (6.3) and (6.4).

Other new results for Pell convolutions emanating from differential equations associated with Gegenbauer polynomials (see [4], [19], and [23]) include

$$\begin{cases} 2k(1 + x^2)P_n^{(k)}(x) = (n + 2k - 1)P_n^{(k-1)}(x) + nxP_{n+1}^{(k-1)}(x) \\ \qquad\qquad = (n + 1)P_{n+2}^{(k-1)}(x) - (n + 2k)xP_{n+1}^{(k-1)}(x) \end{cases}$$
$$(8.4)$$

and

$$4k(k+1)(1+x^2)P_{n-1}^{(k+1)}(x) + 2k(2k+1)xP_n^{(k)}(x) - $$
$$- n(n+2k)P_{n+1}^{(k-1)}(x) = 0. \qquad (8.5)$$

Before reaching (8.5), we derive the intermediate result

$$(1+x^2)\frac{d^2}{dx^2}P_{n+1}^{(k+1)}(x) + (2k+1)x\frac{d}{dx}P_{n+1}^{(k-1)}(x) - $$
$$- n(n+2k)P_{n+1}^{(k-1)}(x) = 0, \qquad (8.6)$$

which is a generalization of the differential equation (3.4) in which $k = 1$.

Recurrence relations for Gegenbauer polynomials may also be employed in the derivation of properties of Pell polynomials.

Can we discover new properties of Gegenbauer polynomials from a further investigation of Pell convolution polynomials? The answer is uncertain.

9. Use of Generalized Humbert Polynomials

In (3.20) the presence of the parameter y allows us to take
$y = -1$ in the case of Pell convolution polynomials and so
avoid, in this specialization from generalized Humbert polynomials, the stricture of a complex variable approach when
using Gegenbauer polynomials.

First, observe from (3.20) and (7.2) that

$$P_{n+1}^{(k)}(x) = P_n(2, x, -1, -(k+1), 1).$$ (9.1)

Then, applying (9.1) to [6, (2.1)] gives the explicit
formula [cf. (7.7)]

$$P_n^{(k)}(x) = \sum_{j=0}^{[(n-1)/2]} \binom{k+j}{j}\binom{k+n-1-j}{n-1-j}(2x)^{n-1-2j}$$ (9.2)

after replacing $n + 1$ by n. Formula (9.2) has [6, (6.2)] the
inversion

$$(2x)^n \binom{k+n}{k} = \sum_{j=0}^{[n/2]} \binom{-k-n-1+j}{j} \times$$

$$\times \frac{k+n+1-2j}{k+n+1-j} P_{n+1-2j}^{(k)}(x).$$ (9.3)

For example, when $n = 3$, $k = 2$, both sides of (9.3) simplify
to $80x^3$ (cf. Table 2).

When $k = 0$, the inversion formula (3.23) for Pell
polynomials ensues.

Application of (9.1) to the recurrence relation [6,
(2.3)] reduces it directly to (7.6).

Coming now to higher-order derivatives w.r.t. x, consider [6, (3.5)]. Simplification, after use of (9.1) with
(3.17) and (3.18), reduces this to

$$\frac{d^k}{dx^k} P_{n+k+1}^{(k)}(x) = 2^k(k+1)!K_k P_{n+1}^{(k)}(x)$$ (9.4)

in terms of the Catalan numbers K_k (3.18). If $n = 0$ in
(9.4), then

$$\frac{d^k}{dx^k} P_{k+1}^{(k)}(x) = 2^k (k+1)! K_k,\tag{9.5}$$

which specializes [6, (3.7)]. [By comparison, the Pincherle polynomial $\mathscr{P}_n(x)$, see (3.24), has k^{th} derivative equal to $(3/2)^k (2k-1)!.$]

Next, let us look at (7.8) again. Differentiate k times w.r.t. x and use (3.11). Put $m = n - 1$. Replace n by $n + 1$. Then set $n = k$. Eventually, *mutatis mutandis*

$$\frac{d^k}{dx^k} P_{k+1}^{(k)}(x) = 2^k (k+1)_k.\tag{9.6}$$

But this is equivalent to (9.5) by (3.19).

10. k^{th} Convolution Sequence $\left\{Q_n^{(k)}(x)\right\}$

Now define the kth *convolution sequence for Pell-Lucas polynomials*

$$\{Q_n^{(k)}(x)\}, \quad n = 1, 2, 3, \ldots$$

by

$$Q_n^{(k)}(x) = \sum_{i=1}^{n} Q_i(x) Q_{n+1-i}^{(k-1)}(x), \quad k \geq 1,$$
$$Q_n^{(0)}(x) = Q_n(x),\tag{10.1}$$

with similar equivalent expressions in (10.1) for $Q_n^{(k)}(x)$ to those in (7.1) for $P_n^{(k)}(x)$. $[Q_0^{(k)}(x) = 0$ if $k \geq 1$; $Q_0^{(0)}(x) = 2.]$

The generating function is

$$\left(\frac{2x + 2y}{1 - 2xy - y^2}\right)^{k+1} = \sum_{n=0}^{\infty} Q_{n+1}^{(k)}(x) y^n.\tag{10.2}$$

To establish an explicit summation formula for $Q_n^{(k)}(x)$, we proceed as follows. From (10.2) and (7.2),

$$\sum_{n=0}^{\infty} Q_{n+1}^{(k)}(x) y^n = \left\{\sum_{n=0}^{k+1} \binom{k+1}{n} (2x)^{k+1-n} 2^n y^n\right\}\left\{\sum_{n=0}^{\infty} P_{n+1}^{(k)}(x) y^n\right\}$$

$$= \sum_{n=0}^{\infty} \left\{\sum_{i=0}^{n} \binom{k+1}{i} (2x)^{k+1-i} 2^i P_{n+1-i}^{(k)}(x)\right\} y^n,$$

whence

$$Q_n^{(k)}(x) = 2^{k+1} \sum_{i=0}^{n-1} \binom{k+1}{i} x^{k+1-i} P_{n-i}^{(k)}(x),\tag{10.3}$$

where $n + 1$ has been replaced by n. Alternatively, we could write

$$\sum_{n=0}^{\infty} Q_{n+1}^{(k)}(x) y^n$$

$$= \sum_{n=0}^{\infty} \left\{ \sum_{i=0}^{n} P_{i+1}^{(k)}(x) \binom{k+1}{n-i} (2x)^{k+1-n+i} 2^{n-i} \right\} y^n,$$

whence

$$Q_n^{(k)}(x) = 2^{k+1} \sum_{i=0}^{n-1} P_{i+1}^{(k)}(x) \binom{k+1}{n-1-i} x^{k+2-n+i} \qquad (10.4)$$

on changing $n + 1$ to n, or, in terms of the complex Gegenbauer polynomial, by (7.10),

$$Q_n^{(k)}(x) = 2^{k+1} \sum_{r=0}^{n-1} (-i)^r \binom{k+1}{n-1-r} C_r^{k+1}(ix) x^{k-n+2+r}.$$

$$(10.5)$$

Both the formulas (10.3) and (10.4) give the summation with emphasis on the subscript n, the number of the polynomial. With emphasis on the superscript k, the number of the convolution, instead, we can derive the alternative formula

$$Q_n^{(k)}(x) = 2^{k+1} \sum_{i=0}^{k+1} \binom{k+1}{i} P_{n-i}^{(k)}(x) x^{k+1-i}, \qquad (10.6)$$

with a corresponding complex Gegenbauer expression as for (10.5).

Formulas (10.3), (10.4), and (10.6) lead to the same result for appropriate values of k and n, of course. For example,

$$Q_3^{(2)}(x) = 192x^5 + 168x^3 + 24x,$$

as in Table 2, by each formula. A more complicated illustration is

$$Q_6^{(3)}(x) = 16[1792x^9 + 3360x^7 + 2040x^5 + 440x^3 + 12x].$$

If we substitute $k = 1$ in (10.6) we have (5.3), whereas when $k = 2$ result (6.13) ensues.

Finally, we comment that it would be nice to be able to exhibit a general explicit formula for $Q_n^{(k)}(x)$ in terms of Pell and Pell-Lucas polynomials only. That is, we desire to generalize (5.5) and (6.13). However, when we aspire to a formula in the case $k = 3$ we obtain, after a little effort, by pursuing the technique utilized for (6.13),

$$Q^{(3)}(x) = \frac{n(n+1)(n+2)}{3} Q_{n+3}(x) - 2n(n+1)P_{n+1}(x) +$$

$$+ 2\left\{\frac{nQ_{n+1}(x) + 2P_n(x)}{4(1+x^2)}\right\}, \tag{10.7}$$

and a possible pattern which appeared to be developing breaks down. Observe that the factor associated with 2 in the last term on the right-hand side of (10.7) is $P_n^{(1)}(x)$, by (4.7).

11. Mixed Pell Convolutions: Conclusion

Much more can be said about convolutions, e.g., convolutions for Pell polynomials with even subscripts. More generally, we can consider *mixed Pell convolutions* involving both $P_n(x)$ and $Q_n(x)$, leading to a *convolution of convolutions*.

But that is another story.

REFERENCES

[1] Bergum, G. E., and Hoggatt, V. E., Jr. 'Limits of Quotients for the Convolved Fibonacci Sequence and Related
Sequences.' *The Fibonacci Quarterly* 15, no. 2 (1977):
113-116.

[2] Byrd, P. F. 'Expansion of Analytic Functions in Polynomials Associated with Fibonacci Numbers.' *The Fibonacci Quarterly* 1, no. 1 (1963):16-29.

[3] Catalan, E. 'Notes sur la théorie des fractions con-
tinues et sur certaines séries.' *Mem. Acad. R. Bel-
gique* 45:1-82.

[4] Erdélyi, A. et al. *Higher Transcendental Functions.*
3 vols. New York: McGraw-Hill, 1953.

[5] Gegenbauer, L. "Zur Theorie der Functionen $C_n^\nu(x)$.'
*Osterreichische Akademie der Wissenschaften Mathema-
tisch Naturwissen Schaftliche Klasse Denkschriften* 48
(1884):293-316.

[6] Gould, H. W. 'Inverse Series Relations and Other
Expansions Involving Humbert Polynomials.' *Duke Math.
J.* 32, no. 4 (1965):697-712.

[7]   Hoggatt, V. E., Jr., and Bicknell-Johnson, Marjorie.
      'Fibonacci Convolution Sequences.' *The Fibonacci
      Quarterly* **15**, no. 2 (1977):117–122.
[8]   Hoggatt, V. E., Jr., and Bicknell-Johnson, Marjorie.
      'Convolution Arrays for Jacobsthal and Fibonacci
      Polynomials.' *The Fibonacci Quarterly* **16**, no. 5
      (1978):385–402.
[9]   Horadam, A. F. 'Pell Identities.' *The Fibonacci
      Quarterly* **9**, no. 3 (1971):245–252, 263.
[10] Horadam, A. F. 'Polynomials Associated with Chebyshev
      Polynomials of the First Kind.' *The Fibonacci Quar-
      terly* **15**, no. 3 (1977):255–257.
[11] Horadam, A. F. 'Chebyshev and Fermat Polynomials for
      Diagonal Functions.' *The Fibonacci Quarterly* **17**, no.
      4 (1979):328–333.
[12] Horadam, A. F. 'Gegenbauer Polynomials Revisited.'
      The Fibonacci Quarterly
[13] Horadam, A. F., and Mahon, J. M., Br. 'Pell and Pell-
      Lucas Polynomials.' *The Fibonacci Quarterly* **23**, no. 1
      (1985):7–20.
[14] Horadam, A. F., and Pethe, S. 'Polynomials Associated
      with Gegenbauer Polynomials.' *The Fibonacci Quarterly*
      19, no. 5 (1981):393–398.
[15] Humbert, P. 'Some Extensions of Pincherle's Polyno-
      mials.' *Proceedings of the Edinburgh Mathematical
      Society* **39**, no. 1 (1921):21–24.
[16] Jacobsthal, E. 'Fibonaccische Polynome und Kreistei-
      lungsgleichungen.' *Berliner Mathematische Gesell-
      schaft. Sitzungsberichte* **17** (1919–1920):43–57.
[17] Lovelock, D. 'On the Recurrence Relation $\Delta_n(z) -
      f(z)\Delta_{n-1}(z) - g(z)\Delta_{n-2}(z) = 0$, and Its Generalisa-
      tions.' *Joint Math. Coll.*, The University of S.A.
      and the University of Witwaterstand, 1967, pp. 139–153.
[18] Lucas, E. *Théôrie des Nombres*. Paris: Blanchard,
      1961.
[19] Magnus, W.; Oberhettinger, F.; and Soni, R. P. *For-
      mulas and Theorems for the Special Functions of Mathe-
      matical Physics*. Berlin: Springer-Verlag, 1966.
[20] Mahon, J. M., Br. M.A. (Hons.) Thesis, University of
      New England, Australia, 1984.
[21] Mahon, J. M., Br., and Horadam, A. F. 'Inverse Trigo-
      nometrical Summation Formulas Involving Pell Polyno-
      mials.' (In press.)

[22]   Pincherle, S.   'Una Nuova Estensione delle Funzioni
       Sferiche.'   *Memorie della R. Accademia di Bologna* 1,
       no 5 (1890):337-369.
[23]   Rainville, E. D.   *Special Functions*.   New York: Mac-
       millan, 1960.
[24]   Rivlin, T. J.   *The Chebyshev Polynomials*.   New York:
       Wiley, 1974.
[25]   Rutherford, D. E.   *Classical Mechanics*.   London:
       Oliver & Boyd, 1957.
[26]   Shannon, A. G.   'Fibonacci Analogs of the Classical
       Polynomials.'   *Math. Mag.* 48, no. 3 (1975):123-130.
[27]   Sneddon, I. N.   *Special Functions of Methematical
       Physics and Chemistry*.   London: Oliver & Boyd, 1956.
[28]   Spraggon, J.   M.A. (Hons.) Thesis, University of New
       England, Australia, 1982.
[29]   Szegö, G.   *Orthogonal Polynomials*.   Amer. Math. Soc.
       Coll. Publications XXIII, 1939.
[30]   Walton, J. E.   M.Sc. Thesis, University of New England,
       Australia, 1968.

A. F. Horadam and A. G. Shannon

CYCLOTOMY-GENERATED POLYNOMIALS
OF FIBONACCI TYPE

1. INTRODUCTION

Fibonacci polynomials may be defined in a number of ways.
For instance, Hoggatt and Bicknell ([4], and elsewhere),
studied Fibonacci polynomials of the form

$$F_{n+2}(x) = xF_{n+1}(x) + F_n(x)$$
$$F_1(x) = 1, \quad F_2(x) = x. \tag{1.1}$$

Roots of (1.1) and related polynomials have been investigated by, e.g., Hoggatt and Bicknell [4], and by Horadam
and Horadam [6].

Kimberling [7] utilized (1.1) in his study of Fibonacci
cyclotomic polynomials to establish some divisibility properties of the Hoggatt-Bicknell polynomials.

Another generalization of Fibonacci polynomials was
developed by Shannon [10], who considered them by analogy
with the classical polynomials of Bernoulli, Euler, and
Hermite. These were quite general because they were based
on the generalized Fibonacci numbers of Horadam [5].

The "Fibonacci cyclotomic polynomials" which we are
about to consider are different from those studied by
Kimberling. They originated from some ideas of Wilson [12],
who generated a number of polynomial equations such as

$$x^3 - x^2 - x = 0 \tag{1.2}$$

and

$$x^7 - 2x^6 - x^4 - x^3 - x^2 - x = 0 \tag{1.3}$$

without specifying the form of the general equation, beyond
stating that its degree is $n^2 + n + 1$, and making some
assertions about its real roots. One of his roots is always
$x = 0$. This zero root will always be neglected by us, and
so we will deal with the corresponding equation of degree
$n^2 + n$.

A. N. Philippou et al. (eds.), Fibonacci Numbers and Their Applications, 81–97.
© *1986 by D. Reidel Publishing Company.*

Wilson [12] further suggested the existence of recurrence sequences of numbers for which these equations would
be characteristic. As his paper is largely of a numerical
nature without any theoretical basis, we propose to clarify,
formalize, and extend his brief treatment.

For subsequent purposes, we record the familiar, but
essential, factual information about Fibonacci numbers in
(1.4)-(1.6) below.

For the Fibonacci sequence of numbers defined by

$$F_{n+2} = F_{n+1} + F_n, \quad F_0 = 0, \ F_1 = 1, \ F_2 = 1, \qquad (1.4)$$

i.e., (1.1) with $x = 1$, the characteristic equation is

$$x^2 - x - 1 = 0, \qquad (1.5)$$

i.e., (1.2) after omitting the common factor.

Roots of (1.5) are

$$\begin{cases} \alpha = \dfrac{1 + \sqrt{5}}{2} \\[3mm] \beta = \dfrac{1 - \sqrt{5}}{2} \end{cases} \quad \begin{array}{l} \text{with } \alpha + \beta = 1, \ \alpha\beta = -1, \\ \alpha - \beta = \sqrt{5}, \ \alpha^2 = \alpha + 1 \qquad (1.6) \\ \text{[from (1.5)]}. \end{array}$$

Equation (1.5) is also characteristic for the sequence
of Lucas numbers defined by

$$L_{n+2} = L_{n+1} + L_n, \quad L_0 = 2, \ L_1 = 1. \qquad (1.7)$$

Acknowledgments are gratefully due to Wilson [12] for
having provided the intriguing and challenging genesis of
our investigation, and to Bowen [1] for valuable contribu-
tions to the theoretical and computational examination of
complex roots in Section 4.

2. CYCLOTOMY-GENERATED POLYNOMIALS
OF FIBONACCI TYPE

Consider the cyclotomic polynomial $\phi_n(x)$, written ϕ_n for
notional convenience, defined as

$$\phi_n = \begin{cases} \dfrac{x^n - 1}{x - 1} & \text{if } x \neq 1 \\ n & \text{if } x = 1 \end{cases} \qquad \text{with } \phi_0 = 0 \qquad (2.1)$$

When $x \neq 1$, (2.1) may of course be expressed as

$$\phi_n = x^{n-1} + x^{n-2} + \cdots + x^2 + x + 1 \qquad (2.2)$$

Information about cyclotomic equations and polynomials may be found, for instance, in Rademacher [8] and Storer [11]. Generally speaking, we will not be much concerned with the use of cyclotomy in any depth.

In terms of cyclotomy, the Wilson polynomials [12] may be expressed generally as

$$\phi_{n^2 + n} = x^{n^2 + n} - x^{2n+1} \frac{\phi_{n^2 - 1}}{\phi_{n+1}} + x^{2n} \frac{\phi_{n^2 - n}}{\phi_n}. \qquad (2.3)$$

Call this polynomial the *cyclotomy-generated polynomial of Fibonacci type* of degree $n^2 + n$. Though cyclotomically generated, these polynomials are not themselves cyclotomic.

The idea of expressing the function in this way, i.e., by (2.3), came from Shannon [9], in which properties of cyclotomic polynomials were studied in the form

$$\mathsf{x}_n = \begin{cases} 1 + x + \cdots + x^{n-1} & (n > 0) \\ 1 & (n = 0) \end{cases} \qquad (2.4)$$

which is also known as the nth *reduced Fermatian of index* x.

Examples of (2.3)

<u>$n = 1$</u> $\phi_2 = x^2$ by (2.3)
                  $= x + 1$ by (2.1),

i.e.,

$$x^2 - x - 1 = 0 \qquad [= (1.5)]. \qquad (2.5)$$

<u>$n = 2$</u> $\phi_6 = x^6 - x^5 + x^4$ by (2.3)

                  $= x^5 + x^4 + x^3 + x^2 + x + 1$ by (2.1),

i.e.,

$$x^6 - 2x^5 - x^3 - x^2 - x - 1 = 0 \quad [\text{cf. } (1.3)]. \tag{2.6}$$

$\underline{n = 3}$
$$\phi_{12} = x^{12} - x^7(x^4 + 1) + x^6(x^3 + 1) \quad \text{by } (2.3), (2.1)$$
$$= x^{11} + x^{10} + x^9 + \cdots + x^2 + x + 1 \quad \text{by } (2.1),$$

i.e.,

$$x^{12} - 2x^{11} - x^{10} - x^8 - 2x^7 - x^5 - x^4 -$$
$$- x^3 - x^2 - x - 1 = 0. \tag{2.7}$$

$\underline{n = 4}$
$$\phi_{20} = x^{20} - x^9(x^{10} + x^5 + 1) + x^8(x^8 + x^4 + 1)$$
$$\text{by } (2.3), (2.1)$$
$$= x^{19} + x^{18} + x^{17} + \cdots + x^2 + x + 1 \quad \text{by } (2.1),$$

i.e.,

$$x^{20} - 2x^{19} - x^{18} - x^{17} - x^{15} - 2x^{14} - x^{13} - x^{11} - x^{10} -$$
$$- 2x^9 - x^7 - x^6 - x^5 - x^4 - x^3 - x^2 - x - 1 = 0. \tag{2.8}$$

Generally,

$$\phi_{n^2+n} = x^{n^2+n} - x^{2n+1}(x^{(n+1)(n-2)} + x^{(n+1)(n-3)} +$$
$$+ x^{(n+1)(n-4)} + \cdots + 1) +$$
$$+ x^{2n}(x^{n(n-2)} + x^{n(n-3)} + x^{n(n-4)} +$$
$$+ \cdots + 1) \quad \text{by } (2.3), (2.1)$$
$$= x^{n^2+n} - (x^{n^2+n-1} + x^{n^2-2} + x^{n^2-n-3} +$$
$$+ \cdots + x^{2n+1}) + (x^{n^2} + x^{n^2-n} + x^{n^2-2n} +$$
$$+ \cdots + x^{2n})$$
$$= x^{n^2+n-1} + x^{n^2+n-2} + x^{n^2+n-3} +$$
$$+ \cdots + x^2 + x + 1 \quad \text{by } (2.1),$$

i.e.,

$$x^{n^2+n} - 2x^{n^2+n-1} - x^{n^2+n-2} - \ldots - x^2 - x - 1 = 0. \tag{2.9}$$

Equations (2.5)-(2.9) are the characteristic equations of the corresponding recurrence relations:

$$\mathscr{I}_{n+2} = \mathscr{I}_{n+1} + \mathscr{I}_n \quad [= (1.4)]; \tag{2.10}$$
$$\mathscr{I}_{n+6} = 2\mathscr{I}_{n+5} + \mathscr{I}_{n+3} + \mathscr{I}_{n+2} + \mathscr{I}_{n+1} + \mathscr{I}_n; \tag{2.11}$$

$$\mathscr{F}_{n+12} = 2\mathscr{F}_{n+11} + \mathscr{F}_{n+10} + \cdots + 2\mathscr{F}_{n+7} +$$
$$+ \cdots + \mathscr{F}_{n+1} + \mathscr{F}_n; \qquad (2.12)$$

$$\mathscr{F}_{n+20} = 2\mathscr{F}_{n+19} + \mathscr{F}_{n+18} + \cdots + 2\mathscr{F}_{n+14} + \mathscr{F}_{n+13} +$$
$$+ \cdots + 2\mathscr{F}_{n+9} + \mathscr{F}_{n+7} + \cdots + \mathscr{F}_{n+1} + \mathscr{F}_n; \qquad (2.13)$$

$$\mathscr{F}_{n^2+n} = 2\mathscr{F}_{n^2+n-1} + \mathscr{F}_{n^2+n-2} + \cdots + \mathscr{F}_{n+1} + \mathscr{F}_n. \qquad (2.14)$$

The pattern of the coefficients 2, 1, 0 of the $\mathscr{F}$'s on the right-hand side of (2.14) can be arranged in n blocks of $n + 1$ digits as follows in Table 1.

Table 1. Pattern of Coefficients in the
         General Recurrence Sequence

$n + 1$

2	1	1	1	... 1	1	0	1
2	1	1	1	... 1	0	1	1
2	1	1	1	... 0	1	1	1
...							
2	1	1	0	... 1	1	1	1
2	1	0	1	... 1	1	1	1
2	0	1	1	... 1	1	1	1
1	1	1	1	... 1	1	1	1

n

Zeros occur as coefficients of $\mathscr{F}_{n^2}$, $\mathscr{F}_{n(n-1)}$, $\mathscr{F}_{n(n-2)}$, $\mathscr{F}_{n(n-3)}$, $\cdots$, $\mathscr{F}_{n(n-(n-2))} = \mathscr{F}_{2n}$.

The last block (row) of Table 1 contains $n + 1$ one's. In (2.14), the last $2n$ coefficients are all equal to 1, a fact which is related to the term in x^{2n} in (2.3).

Of course, the coefficients appearing in Table 1 are also the negatives of the $n^2 + n$ coefficients in (2.9) occurring after the first term.

Each of the recurrence relations (2.10)-(2.14) of order 2, 6, 12, 20, $\ldots$, $n^2 + n$ defines an infinite set of sequences depending on the initial set of 2, 6, 12, 20, $\ldots$, $n^2 + n$ numbers specified.

Wilson [12] chose certain representative sequences to illustrate the principles being formulated and it is these that concern us. For small values of n, the first few of these *associated sequences* are tabulated in Table 2 for

Table 2. Associated Sequences (n = 1, 2, ..., 6)

r	... -4 -3 -2 -1 0 1 2 3 4 ...
... -8 5 -3 2 -1 1 0 1 1 2 3 5 8 ...
... -17 11 -8 6 -4 3 -2 1 -1 1 0 1 1 3 7 17 39 ...
... -14 11 -9 7 -6 5 -4 3 -2 2 -1 1 -1 1 0 3 5 15 36 92 231 ...
... -7 6 -5 4 -3 3 -2 2 -2 1 -1 1 -1 1 0 2 4 11 28 73 188 485 ...
... -14 12 -10 9 -8 7 -6 5 -4 4 -3 3 -3 2 -2 1 -1 1 -1 1 0 3 6 16 40 105 274 712 1856 4833 12589 32787 ...
... -10 9 -8 7 -6 5 -4 3 -3 2 2 -2 2 -1 1 -1 1 -1 1 -1 1 0 2 5 13 34 89 231 610 1591 4155 10858 23870 ...

visual effect, where r refers to the subscript of the number $\mathscr{F}_r$.

Observe the occurrence for $r \geqslant 0$ of:

($n = 1$) the Fibonacci sequence;

($n = 6$) the second, third, . . ., sixth terms of the
Fibonacci sequence of odd subscripts with the
first term missing. Wilson [12] comments that
"later terms only slowly tend to fall below
second Fibonacci terms."

3. REAL ROOTS

In the discussion that follows, one might be guided by reference to the case $n = 4$, i.e., to (2.8) and (2.13).

Employing Descartes's Rule of Signs [2], we see that there is, at most, one positive root for each equation (2.9) as n varies.

Computer calculations show that there is *exactly* one positive root for each n. Table 3 gives approximate values of these roots to five decimal places for some small values of n.

Table 3. Values of Positive and Negative Roots

	x_{1n}	x_{2n}
$n = 1$	1.61803	−0.61803
2	2.31651	−0.69848
3	2.51650	−0.80459
4	2.58161	−0.82780
5	2.60453	−0.86853
6	2.61295	−0.87944
7	2.61610	−0.90093
8	2.61730	−0.90724
9	2.61776	−0.92051
10	2.61793	−0.92463
⋮	⋮	⋮
20	2.61803	−0.96108

Designate these positive roots by x_{1n} $(n = 1, 2, 3, \ldots)$.

Conjecture 1: There is exactly one negative root for each n.

Denote these negative roots by x_{2n} $(n = 1, 2, 3, \ldots)$. Values of x_{2n} for small n appear in Table 3 above. They consolidate the validity of the conjecture.

From these numerical data, it appears that

$$\alpha \leqslant x_{1n} < 1 + \alpha \ \left(= 2.61803\ 39887 \ldots = \frac{3 + \sqrt{5}}{2} = \alpha^2\right)$$

and

$$-1 < x_{2n} \leqslant -\frac{1}{\alpha} \ (= \beta).$$

[See (1.6).] The negative root approaches its lower bound more slowly than the positive root approaches its upper bound.

Because of the near certainty of the validity of the conjectures, which we assume to have the force of theorems, we shall number conjectures and theorems consecutively without distinguishing their different natures.

Theorem 2: $\lim\limits_{n \to \infty} x_{1n} = \alpha^2$.

Proof: Rewrite (2.3) in cyclotomic form:

$$\frac{x^{n^2 + n} - 1}{x - 1} = x^{n^2 + n} - x^{2n+1}\left(\frac{x^{n^2 - 1} - 1}{x^{n+1} - 1}\right) +$$

$$+ \ x^{2n}\left(\frac{x^{n^2 - n} - 1}{x^n - 1}\right) \cdots \text{(i)}. \qquad\qquad (2.3)'$$

Divide throughout by $x^{n^2 + n}$.
Then let $n \to \infty$.
For $x > 1$, we have, from (i),

$$\frac{1}{x - 1} = 1 - \frac{1}{x},$$

i.e.,

$$x^2 - 3x + 1 = 0. \qquad\qquad (3.1)$$

Roots of (3.1) are

$$\begin{cases} x_1 = \dfrac{3 + \sqrt{5}}{2} = \alpha^2 = 1 + \alpha \quad (>1) \\[4mm] x_1' = \dfrac{3 - \sqrt{5}}{2} = \beta^2 = 2 - \alpha \quad (<1) \end{cases} \quad \text{by (1.6).} \quad (3.2)$$

Discard the root less than unity.
Then $x_1 = \alpha^2$ is the required limit.

Comments on the Theorem

(1) Later, in Section 4, we establish another approach to the proof of this Theorem.

(2) Equation (3.1) is precisely the characteristic equation of the sequence of odd-subscripted Fibonacci numbers (see Catalan [3]), i.e., of the sequence 1, 2, 5, 13, 34, 89, ... It is also the characteristic equation for even-subscripted Fibonacci numbers. (Historically, it is interesting to observe that the Fibonacci sequence is also called by Catalan the *Lamé sequence*.)

Conjecture 3: $\lim\limits_{n \to \infty} x_{2n} = -1.$

More will be said about this conjectural assertion in the next section.

Taken together, Theorem 2 and Conjecture 3 imply that, for n large, there is a component of (2.9) which approaches $(x + 1)(x - \alpha^2) = x^2 - \alpha x - \alpha^2$. Otherwise expressed,

$$\lim_{n \to \infty}(x_{1n} + x_{2n}) = \alpha,$$

$$\lim_{n \to \infty}(x_{1n} x_{2n}) = -\alpha^2.$$

4. COMPLEX ROOTS

Now, in accordance with our previous convention for ϕ_n, let us denote the polynomial (2.9) by p_n where here, and in polynomials about to be introduced, we omit the functional notation, i.e., $p_n = p_n(x)$.

Of the $n^2 + n$ roots of equation (2.9), i.e., $p_n = 0$, only two are real. One of these is positive (Theorem 2), the other negative (Conjecture 1). Bowen [1] has conjectured that the remaining $n^2 + n - 2$ roots of $p_n = 0$, which are complex, are geometrically located in the complex plane in two categories.

Conjecture 4: $p_n = 0$ has $n^2 - n$ complex roots which lie *on* the unit circle $|x| = 1$ and are of the form $\omega^{\pm k}$, where $\omega = e^{2\pi i/n(n+1)}$ and k is a positive integer.

$$\left(k \neq sn, \ s = 1, 2, \ldots, \left[\frac{n}{2}\right]; \right.$$
$$\left. k \neq t(n+1), \ t = 1, 2, \ldots, \left[\frac{n-1}{2}\right]. \right)$$

Conjecture 5: $p_n = 0$ has $2n - 2$ complex roots which lie *within* the unit circle.

Example $(n = 4)$: Apart from the two real roots (Table 3), the eighteen complex roots are as shown in Table 4 (to five decimal places).

Table 4. Location of Complex Roots $(n = 4)$

	Root	Modulus	(±)Argument	k
12 on the	.95115 ± .30891i	1	18°	1
unit circle	.80925 ± .58773i	1	36°	2
	.58809 ± .80906i	1	54°	3
	−.30923 ± .95113i	1	108°	6
	−.58794 ± .80908i	1	126°	7
	−.95107 ± .30897i	1	162°	9
6 inside the	.35911 ± .85097i	.92364	67.120°	4
unit circle	−.06789 ± .86805i	.87070	94.472°	5
	−.66837 ± .52791i	.85171	141.696°	8

Missing from the set of 12 are the complex roots with arguments 72°, 90°, 144° (and 180°) corresponding to the values $k = 4, 5, 8$ (10). These missing arguments bear some resemblance to the arguments of the set of 6 inside the unit circle, i.e., for which $|x| < 1$. Numerical tabulation for the case $n = 5$ (suppressed here) provides us with correspondingly similar details which reinforce the statements

CYCLOTOMY-GENERATED POLYNOMIALS OF FIBONACCI TYPE   91

for $n = 4$. Moreover, for $n = 5$, the moduli of the eight
points within the unit circle show the tendency to move
closer to 1, which is evident as we increase n from 2,
through 3, to 4.
    Let

$$q_n = \prod_k (x - \omega^k)(x - \omega^{-k}),$$ (4.1)

i.e., q_n is of degree $n^2 - n$.
    Thus,

$$q_n = (p_n, \phi_{n^2 + n}).$$ (4.2)

    Then this greatest common divisor q_n is such that

$$p_n = q_n r_n,$$ (4.3)

where r_n is of degree $2n$, and contains the two real roots.

Examples of (4.3)

$\underline{n = 2}$ $\begin{cases} q_2 = x^2 - x + 1 \\ r_2 = x^4 - x^3 - 2x^2 - 2x - 1 \end{cases}$

$\underline{n = 3}$ $\begin{cases} q_3 = x^6 - x^5 + x^3 - x + 1 = q_2(x^4 - x^2 + 1) \\ r_3 = x^6 - x^5 - 2x^4 - 3x^3 - 3x^2 - 2x - 1 \end{cases}$

$\underline{n = 4}$ $\begin{cases} q_4 = x^{12} - x^{11} + x^8 - x^6 + x^4 - x + 1 \\ r_4 = x^8 - x^7 - 2x^6 - 3x^5 - 4x^4 - 4x^3 - 3x^2 - 2x - 1 \end{cases}$

    One can discern a pattern of coefficients emerging for
q_n and r_n.
    If we write only the coefficients of the powers of r_n,
we have the pattern of $2n + 1$ numbers

$$\overbrace{1 \ -1 \ -2 \ -3 \ -4 \dots -(n-1) \ -n}^{} \ \overbrace{-n \ -(n-1) \dots -3 \ -2 \ -1}^{}$$
(4.4)

    Multiply the polynomial r_n of degree $2n$ respresented by
(4.4) by $(1 - x)^2$.
    Simplification reduces the expression to a six-term
polynomial r_n^* of degree $2n + 2$, which, when equated to zero,
becomes

$$x^{2n+2} - 3x^{2n+1} + x^{2n} + x^{n+1} + x^n - 1 = 0, \qquad (4.5)$$

wherein $x = 1$ is a double root. A virtue of this equation is that it provides easier access to a study of the real roots of (2.9).

For large n and $x > 1$, equation (4.5) is dominated by the first three terms, i.e., by

$$x^2 - 3x + 1.$$

As an equation, this is precisely (3.1) whose positive root > 1 is α^2. This affords us another viewpoint for the proof of Theorem 2 (i.e., $\lim_{n \to \infty} x_{1n} = \alpha^2$).

Descartes's Rule of Signs [2] applied to (4.4), rather than (4.5), shows us immediately that there is just *one* negative root. Thus,

Conjecture 1 is true if (4.4) is true

(i.e., if Conjectures 4 and 5 are true).
 Now,

$$\left.\begin{array}{l} r_n(0) = -1 \\[2mm] r_n(-1) = 1 \end{array}\right\} \text{ from (4.4),}$$

i.e., the negative root is located between 0 and -1, as we have earlier seen empirically.

Regarding the pattern of coefficients for q_n [cf. (4.4) for the pattern for r_n], we find that an analysis of the situation shows a relationship to the corresponding pattern in Table 1 relating to p_n to be given in the following table (Table 5), where we omit the first term x^{n^2-n} and the last two, $-x$ and $+1$.

Table 5. Pattern of Coefficients in q_n

	$n + 1$								
	-1	0	0	0	... 0	0	1	0	
	-1	0	0	0	... 0	1	0	0	
	-1	0	0	0	... 1	0	0	0	
$n - 2$	....
	-1	0	0	1	... 0	0	0	0	
	-1	0	1	0	... 0	0	0	0	

This pattern bears a close resemblance to that in the first $n - 2$ rows of Table 1 if we replace 2, 1, 0 therein by -1, 0, 1, respectively. One readily checks that the degree of r_n is $(n - 2)(n + 1) + 1 + 1 = n^2 - n$.

We have thus established the general forms of p_n, q_n, and r_n for any integer n ($\geqslant 1$).

Two further remarks may be made here:

(i) Algebraic manipulation of (4.5) can lead to

$$x^{2n} = \phi_{n+1}\phi_n$$

after division by $(1 - x)^2$.

(ii) Temporarily denoting the right-hand side of (2.3) by $\phi^*_{n^2+n}$, we may discover that

$$\phi^*_{n^2+n} = x^{2n} q_n.$$

5. POLYNOMIAL CURVES

It is of interest to have a quick look at some of the polynomial curves $y = p_n(x)$ for small values of n.

First, observe the common features of these curves for all n:

$$p_n(0) = -1; \tag{5.1}$$

$$p_n'(0) = -1; \tag{5.2}$$

i.e., the infinite family of polynomial curves have a common tangent with gradient 135° at the point $(0, -1)$.

Figure 1 gives the salient characteristics of the computer-drawn curves $y = p_n(x)$, where $n = 1, 2, 3, 4$, in the domain bounded by the real roots. In particular, the curves are shown to possess just one real root and just one negative root. Observe, where it is visible, the existence of an inflection (or inflections), except in the case of the parabola ($n = 2$), of course.

The different scales used for the graphs of $y = p_n(x)$, $n = 1, 2, 3, 4$, should be noted. The large vertical scales necessary for the cases $n = 3, 4$ render it impossible to witness the behavior of the curves in the vicinity of the origin.

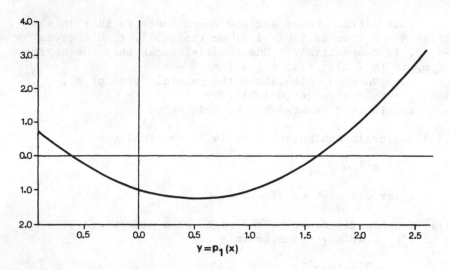

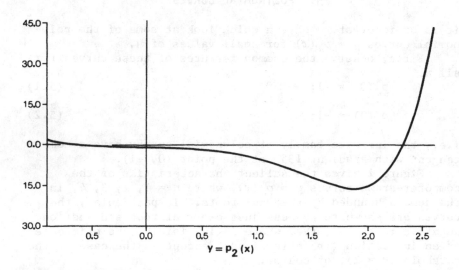

Figure 1. Polynomial Curves $y = p_n(x)$ $(n = 1, 2, 3, 4)$

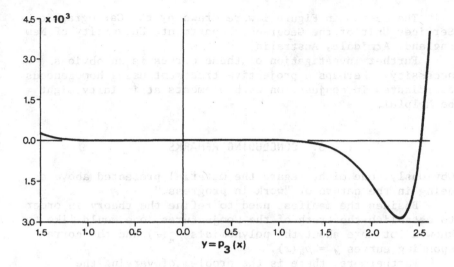

$$y = P_3(x)$$

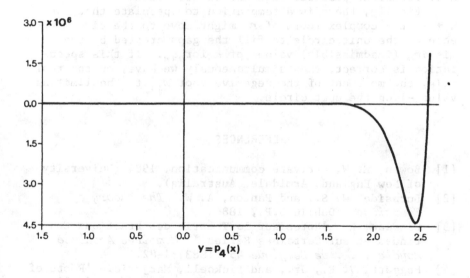

$$y = P_4(x)$$

Figure 1 (continued)

The curves in Figure 1 were drawn by the Cartographic
Services Unit of the Geography Department, University of New
England, Armidale, Australia.

Further investigation of these curves is an obvious
necessity. Perhaps a projective treatment using homogeneous
coordinates in conjunction with elements at infinity might
be helpful.

6.   CONCLUDING REMARKS

Obviously, one might regard the material presented above as
being in the nature of "work in progress."

Besides the manifest need to refine the theory in order
to establish the truth of the Conjectures, one would like to
know a lot more about the polynomials $p_n(x)$ and the corresponding curves $y = p_n(x)$.

Furthermore, there is the problem of varying the
sequence, for each value of n, from the one used here which
Wilson [12] called "atypical because they are close to the
most efficient possible sequence in each case."

Finally, there is a temptation to speculate that, as
$n \to \infty$, the complex roots of r might move to the circumference of the unit circle to fill the gaps created by the
missing (inadmissible) values of k for q_n. If this specu-
lation is correct, then simultaneously we have, on the real
axis, the movement of the negative root x_{1n} to the limiting
value -1 on the unit circle.

REFERENCES

[1]   Bowen, E. W.   Private communication, 1984 (University
       of New England, Armidale, Australia).
[2]   Burnside, W. S., and Panton, A. W. *The Theory of
       Equations.* Dublin U.P., 1886.
[3]   Catalan, E.   'Notes sur la Théorie des Fractions Con-
       tinues et sur Certaines Séries.' *Memoires Academie
       Royale . . . de Belgique* 45 (1883):1-82.
[4]   Hoggatt, V. E., Jr., and Bicknell, Marjorie.   'Roots of
       Fibonacci Polynomials.' *The Fibonacci Quarterly* 11,
       no. 3 (1973):271-274.
[5]   Horadam, A. F.   'A Generalized Fibonacci Sequence.'
       American Mathematical Monthly 68, no. 5 (1961):455-459.

[6]   Horadam, A. F., and Horadam, E. M.   'Roots of Recur-
      rence-Generated Polynomials.'   *The Fibonacci Quarterly*
      20, no. 3 (1982):219-226.
[7]   Kimberling, C.   'Generalized Cyclotomic Polynomials,
      Fibonacci Cyclotomic Polynomials, and Lucas Cyclotomic
      Polynomials.'   *The Fibonacci Quarterly* **18**, no. 2
      (1980):108-126.
[8]   Rademacher, H.   *Lectures on Elementary Number Theory.*
      London: Blaisdell, 1964.
[9]   Shannon, A. G.   'Applications of Some Techniques of
      L. Carlitz to Some Recurrence Relations.'   M.A. Thesis,
      University of New England, Armidale, Australia, 1972.
[10] Shannon, A. G.   'Fibonacci Analogs of the Classical
      Polynomials.'   *Mathematics Magazine* **48**, no. 3 (1975):
      123-130.
[11] Storer, T.   *Cyclotomy and Difference Sets.*   London:
      Markham, 1967.
[12] Wilson, L. G.   *Other Fibonacci Type Sequences.*   Pri-
      vately published, Brisbane, Australia, 1979.

Daniela Jaruškova

ON GENERALIZED FIBONACCI PROCESS

Let us consider a discrete-time population model in which an individual entering the population in year t gives birth in years $\{t + i + 1, i < k\}$ to one offspring i-times a year and in years $\{t + i + 1, i \geqslant k\}$ to one offspring k-times a year. If we denote Y_n^i $(i = 0, 1, \ldots, k - 1)$ the number of the individuals which are in the population i years and Y_n^k the number of the individuals which are in the population k or more than k years, then the sequence $Y_n = (Y_n^0, \ldots, Y_n^k)$ satisfies the following equations:

$$Y_{n+1}^0 = Y_n^1 + 2Y_n^2 + \cdots + kY_n^k,$$

$$Y_{n+1}^1 = Y_n^0,$$

$$Y_{n+1}^2 = Y_n^1,$$

$$\vdots$$

$$Y_{n+1}^k = Y_n^{k-1} + Y_n^k,$$

and the numbers of all individuals

$$X_n = \sum_{j=0}^{k} Y_n^j$$

form a Fibonacci sequence of order k,

$$X_{n+1} = X_n + X_{n-1} + \cdots + X_{n-k}.$$

If we suppose the random case where the numbers of offsprings produced by the ith individual in the year n at the jth throw $\{Z_{ji}^n\}$ are i.i.d. with $EZ_{ji}^n = m$, var $Z_{ji}^n = \sigma^2$, $E(Z_{ji} - m)^4 = d < \infty$, then the total number of individuals X_n forms a generalized Fibonacci branching process

$$X_n = X_{n-1} + \sum_{i=1}^{X_{n-2}} Z_{1i}^n + \cdots + \sum_{i=1}^{X_{n-k-1}} Z_{ki}^n. \tag{1}$$

99

A. N. Philippou et al. (eds.), Fibonacci Numbers and Their Applications, 99–104.
© *1986 by D. Reidel Publishing Company.*

Further, we may use the well-known results of the theory of the multitype Galton–Watson process where the matrix

$$M = \|m_{ij}\|_{i,j=0}^{k} \quad (m_{ij} = EY_i^j/Y_0 = e_i)$$

and its powers M^n play an important role. The behavior of the matrices M^n is determined by ρ, the greatest eigenvalue of M. The value ρ ($\rho > 1$) is called the growth rate because Y_n/ρ^n converge with probability 1 to $W\nu_\rho$, where W is a random variable and ν_ρ is the left eigenvector corresponding to ρ. In our case the matrix M has the form

$$M = \begin{Vmatrix} 0 & 1 & 0 & 0 & \dots & 0 \\ m & 0 & 1 & 0 & \dots & 0 \\ 2m & 0 & 0 & 1 & \dots & 0 \\ \vdots & & & & & \\ (k-1)m & 0 & 0 & 0 & \dots & 1 \\ km & 0 & 0 & 0 & \dots & 1 \end{Vmatrix},$$

with the characteristic polynomial $Q_{k+1}(x)$ where, for $n \geqslant 1$, $Q_n(x) = x^n - x^{n-1} - mx^{n-2} - \cdots - m$.

The most simple estimator of ρ is the ratio estimator $\hat{\rho} = X_N/X_{N-1}$, which is strongly consistent in our case. It is interesting that the rate of convergence X_N/X_{N-1} depends on the relation between $z_1 = \rho$ and λ, where λ is the second maximal absolute value of the eigenvalues of the matrix M. We can distinguish three cases: $\lambda^2 < \rho$, $\lambda^2 = \rho$, and $\lambda^2 > \rho$.

Let us investigate the behavior of the differences $D_N = X_N - \rho X_{N-1}$. If we introduce the random variables $U_n = X_n - X_{n-1} - mX_{n-2} - \cdots - mX_{n-k-1}$, which are in consequence of (1) the martingale differences w.r.t. σ-fields generated by $X_0, X_1, \ldots, X_n$, we can write

$$D_N = U_N - (\rho - 1)X_{N-1} + mX_{N-2} + \cdots + mX_{N-k-1}$$

$$= U_N - (\rho - 1)D_{N-1} - (\rho - 1)\rho X_{N-2} +$$

$$+ mX_{N-2} + \cdots + mX_{N-k-1}$$

$$= U_N - Q_1(\rho)D_{N-1} - Q_2(\rho)D_{N-2} -$$

$$- Q_3(\rho)D_{N-3} - \cdots - Q_k(\rho)D_{N-k}.$$

This gives, on iteration,

$$D_N = \alpha_0 U_N + \alpha_1 U_{N-1} + \cdots + \alpha_{N-k-1} U_{k+1} +$$
$$+ \beta_N^k D_k + \cdots + \beta_N^1 D_1,$$

where the coefficients $\{\alpha_n\}$ and $\{\beta_n^j, \ j = 1, \ \ldots, \ k\}$ satisfy the following system of difference equations:

$$\alpha_n + Q_1(\rho)\alpha_{n-1} + \cdots + Q_k(\rho)\alpha_{n-k} = 0,$$

$$\beta_n^j + Q_1(\rho)\beta_{n-1}^j + \cdots + Q_k(\rho)\beta_{n-k}^j = 0,$$

for $j = 1, \ \ldots, \ k.$

The behavior of $\{\alpha_n\}$ and $\{\beta_n^j\}$ is given by the roots of the characteristic polynomial $R(x) = x^k + Q_1(\rho)x^{k-1} + \cdots + Q_k(\rho)$. Using the fact that $Q_{n+1}(x) = xQ_n(x) - m$, it is easy to show that if the polynomial $Q_{k+1}(x)$ has the roots $z_1 = \rho, \ z_2, \ \ldots,$ z_{k+1}, then the polynomial $R(x)$ has the roots $z_2, \ \ldots, \ z_{k+1}$. Consequently, the second maximal absolute value of the roots of $Q_{k+1}(x)$ is the maximal absolute value of all roots of $R(x)$. Generally, the polynomial $Q_{k+1}(x)$ can have more roots with the second maximal absolute value λ. The asymptotic behavior of α_n and β_n^j is given by the roots with the absolute value λ and with the maximal multiplicity γ. Especially, if we denote these roots $\lambda \exp(i\varphi_1), \ \ldots, \ \lambda \exp(i\varphi_r)$, we can write

$$\alpha_n = \lambda^n n^{\gamma-1}(C_1 \exp(in\varphi_1) + \cdots + C_r \exp(in\varphi_r)) +$$

$$+ \ o(\lambda^n n^{\gamma-1})$$

$$\beta_n^j = \lambda^n n^{\gamma-1}(C_1^j \exp(in\varphi_1) + \cdots + C_r^j \exp(in\varphi_r)) +$$

$$+ \ o(\lambda^n n^{\gamma-1}), \ j = 1, \ \ldots, \ k. \tag{2}$$

Let us introduce the variables T_N:

$$T_N = \alpha_0 U_N + \cdots + \alpha_{N-k-1} U_{k+1}$$

$$= \alpha_0 \left((Z_{11}^N - m) + \cdots + (Z_{1X_{N-2}}^N - m) + \cdots + \right.$$

$$+ \ (Z_{kX_{N-1-k}}^N - m) \right) + \cdots + \alpha_{N-k-1} \left((Z_{11}^{k+1} - m) + \right.$$

$$+ \ \cdots + (Z_{kX_0}^{k+1} - m) \right).$$

CASE 1: $\lambda^2 < \rho$

The variable $(\hat{\rho} - \rho)\rho^{N/2}W^{1/2}$ is asymptotically normally distributed with zero mean and finite variance.

Proof: The assertion is an easy conclusion of the fact that $D_N/\rho^{N/2}W^{1/2}$ is asymptotically normal, which will be proved. It holds that $E(\beta_N^k D_k + \cdots + \beta^1 D_1)^2/\rho^N \to 0$. For the expression $T_N/\rho^{N/2}$, the central limit theorem can be used.

Let us enumerate all throws and introduce σ-fields $\mathcal{G}_j = \sigma a(z_0, \ldots, z_j)$, where z_j denotes the number of offsprings born at the jth throw. Then

$$V_{Nj} = \frac{z_j - m}{\rho^{N/2}}\alpha_{N-n(j)} \quad \text{for } n(j) \leqslant N$$

$$= 0 \quad \text{for } n(j) > N,$$

where the expression $n(j)$ denotes the year in which the throw number j occurred, are martingale differences w.r.t. $\{\mathcal{G}_j\}$ and $T_N/\rho^{N/2} = \sum_{j=0}^{\infty} V_{Nj}$. Now we can use the central limit theorem which asserts that, under the conditions

(a) $E \sum_{j=0}^{\infty} E(V_{Nj}^2/\mathcal{G}_{j-1}) < \infty$,

(b) $S_N^2 = \sum_{j=0}^{\infty} E(V_{Nj}^2/\mathcal{G}_{j-1}) \xrightarrow{P} W$ constant,

(c) $d_N^4 = \sum_{j=0}^{\infty} E(V_{Nj}^4/\mathcal{G}_{j-1}) \xrightarrow{P} 0$,

the limit distribution of $W^{-1/2}\sum_{j=0}^{\infty} V_{Nj}$ is normal. The conditions (a)-(c) for

$$s_N^2 = \frac{\sigma^2}{\rho^N} \sum_{n=k+1}^{N} \alpha_{N-n}^2 \sum_{e=1}^{k} X_{n-1-e}$$

and

$$d_N^4 = \frac{d}{\rho^{2N}} \sum_{n=k+1}^{N} a_{N-n}^4 \sum_{e=1}^{k} X_{n-1-e}$$

are satisfied because of $X_n/\rho^n \to W$ and $\lambda^2 < \rho$.

CASE 2: $\lambda^2 = \rho$

The variable $(\hat{\rho} - \rho)\rho^{N/2}W^{1/2}/N^{\gamma-1/2}$ is asymptotically normally distributed with zero mean and finte variance.

Proof: The proof of Case 2 is analogous to the proof of the preceding assertion.

CASE 3: $\lambda^2 > \rho$

There exist random variables $A_1, \ldots, A_r$ where

$$A_p = C_p \sum_{j=k+1}^{\infty} \exp(-ij\varphi_p)\frac{U_j}{\lambda^j} + C_p^k D_k +$$

$$+ \cdots + C_p^1 D_1, \quad p = 1, \ldots, r,$$

such that

$$\left| \frac{(\hat{\rho} - \rho)\rho^{N-1}W}{N^{\gamma-1}\lambda^N} - \sum_{p=1}^{r} \exp(iN\varphi_p)A_p \right| \to 0 \quad \text{a.s.}$$

Proof: First we note that

$$EU_n^2 = E\, E(U_n^2/X_0, \ldots, X_{n-1})$$

$$= E\Big(E(Z_{11}^n - m)^2 + \cdots + (Z_{1X_{n-2}}^n - m)^2 +$$

$$+ \cdots + (Z_{kX_{n-k-1}}^n - m)^2\Big)$$

$$= E(\sigma^2 X_{n-2} + \cdots + \sigma^2 X_{n-k-1}).$$

The sequence $\{EX_n\}$ satisfies the difference equation

$$EX_n = EX_{n-1} + mEX_{n-2} + \cdots + mEX_{n-k-1},$$

with the characteristic polynomial $Q_{k+1}(x)$. Consequently, $EX_n = K\rho^n + o(\rho^n)$ and $EU_n^2 \leqslant$ constant ρ^n. Further, using (2), we can write

$$\frac{D_N}{N^{\gamma-1}\lambda^N} = \sum_{p=1}^{r} \exp(iN\varphi_p)\Big(C_p \sum_{j=k+1}^{N} \exp(-ij\varphi_p) \times$$

$$\times \frac{(N-j)^{\gamma-1}}{N^{\gamma-1}} \frac{U_j}{\lambda^j} + C_p^k D_k + \cdots + C_p^1 D_1\Big) + R_N$$

where $R_N \to 0$ a.s. and for $p = 1, \ldots, r$,

$$\sum_{j = k+1}^{N} \exp(-ij\varphi_p) \frac{(N - j)^{\gamma-1}}{N^{\gamma-1}} \frac{U_j}{\lambda^j} \to$$

$$\to \sum_{j = k+1}^{\infty} \exp(-ij\varphi_p) \frac{U_j}{\lambda^j} \quad \text{a.s.}$$

as a consequence of the fact that

$$\sum_{j = k+1}^{\infty} \frac{|U_j|}{\lambda^j} < \infty \quad \text{a.s.}$$

REFERENCES

Asmussen, S., and Keiding, N. 'Martingale Central Limit
    Theorems and Asymptotic Estimation Theory for Multitype
    Branching Processes.' *Adv. Appl. Prob.* **10** (1978):109-
    129.

Asmussen, S. 'On the Role of a Certain Eigenvalue in Esti-
    mating the Growth Rate of a Branching Process.' Pre-
    print January 1981, No. 2, Institute of Mathematical
    Statistics, University of Copenhagen.

Heyde, C. C. 'On Fibonacci (or Lagged Bienayme-Galton-
    Watson) Branching Processes.' *J. Appl. Prob.* **18** (1981):
    583-591.

Kesten, H., and Stigum, B. P. 'Additional Limit Theorems
    for Indecomposable Multidimensional Galton-Watson
    Processes.' *Ann. Math. Statist.* **37** (1966):1463-1481.

Peter Kirschenhofer, Helmut Prodinger,
and Robert F. Tichy

FIBONACCI NUMBERS OF GRAPHS III:
PLANTED PLANE TREES

1. INTRODUCTION

In [8], the Fibonacci number $f(G)$ of a (simple) graph G is introduced as the total number of all Fibonacci subsets S of the vertex $V(G)$ of G, where a Fibonacci subset S is a (possibly empty) subset of $V(G)$ such that any two vertices of S are not adjacent. In Graph Theory, a Fibonacci subset is called an independent or internally stable set of vertices.

In [6], the average Fibonacci number of binary trees of size n has been considered for the first time: The family $\mathscr{B}$ of all binary trees is defined by the following equation ($\square$ is the symbol for a leaf and $\bigcirc$ for an internal node), compare [7]:

$$\mathscr{B} = \square + \overset{\bigcirc}{\underset{\mathscr{B} \quad \mathscr{B}}{\diagup\diagdown}} . \qquad (1.1)$$

Denoting by $h_n(\mathscr{B})$ the total number of Fibonacci subsets of all binary trees of size n, i.e., with n internal nodes, it has been shown in [6] that

$$h_n(\mathscr{B}) \sim (0.63713\ldots)(0.15268\ldots)^{-n} \cdot n^{-3/2} , \qquad (1.2)$$

so that the average value $S_n(\mathscr{B})$ of the Fibonacci number of a binary tree of size n fulfils asymptotically

$$S_n(\mathscr{B}) \sim (1.12928\ldots)(1.63742\ldots)^n . \qquad (1.3)$$

In Section 2 of the present paper we present an explicit formula for $h_n(\mathscr{B})$ and exact values for the numerical constants in (1.2) and (1.3).

In Section 3 we generalize the foregoing results to the family $\mathscr{T}$ of t-ary trees. Moreover, we determine how the number $h_n(\mathscr{T})$ of all Fibonacci subsets divides up into the numbers $h_{n,j}(\mathscr{T})$ of Fibonacci subsets of cardinality j. The

A. N. Philippou et al. (eds.), Fibonacci Numbers and Their Applications, 105–120.

order of exponential growth of $S_n(\mathcal{F})$ is analyzed asymptotically for $t \to \infty$.

In the last section we deal with the family $\mathcal{P}$ of all planted plane trees. As a paradigm, we also investigate the second-order moments.

2.   BINARY TREES

Let $f_n = f_n(\mathcal{B})$ resp. $g_n = g_n(\mathcal{B})$ be defined by

$$f_n = \sum_{T \in \mathcal{B}_n} \text{card}\{S : S \subseteq V(T); \ S \text{ a Fibonacci subset } not \text{ containing the root}\},$$

$$g_n = \sum_{T \in \mathcal{B}_n} \text{card}\{S : S \subseteq V(T); \ S \text{ a Fibonacci subset containing the root}\},$$

where $\mathcal{B}_n$ denotes the family of trees of size n in $\mathcal{B}$. Also let

$$f(z) = \sum_{n \geqslant 1} f_n z^n \quad \text{and} \quad g(z) = \sum_{n \geqslant 1} g_n z^n$$

be the corresponding generating functions. From (1.1), we derive the functional equations

$$f = z(1 + f + g)^2$$
$$g = z(1 + f)^2. \tag{2.1}$$

Substituting $u = z(1 + f)$, we have

$$g = u^2/z, \quad f = u^2(1 + u)^2/z,$$

whence

$$u = z + zf = z + u^2(1 + u)^2 \tag{2.2}$$

or

$$z = \frac{u}{\varphi(u)} \quad \text{with} \quad \varphi(u) = (1 - u(1 + u)^2)^{-1}.$$

Applying Lagrange's inversion formula (LIF) (see [3]), it turns out that the coefficients $[z^n]u(z)$ of z^n in the (formal) power series $u(z)$ fulfil

$$[z^{n+1}]u(z) = \frac{1}{n+1}[w^n](1 - w(1 + w)^2)^{-n-1}$$
$$= \frac{1}{n+1}\sum_{j=0}^{n}\binom{n+j}{j}\binom{2j}{n-j}, \tag{2.3}$$

from which the coefficients of $f(z)$ are immediate. In order to expand $u^2 = zg$, we again apply LIF to get

$$[z^{n+1}]u^2(z) = \frac{2}{n+1}[w^{n-1}](1 - w(1 + w)^2)^{-n-1}$$

$$= \frac{2}{n+1}\sum_{j=0}^{n-1}\binom{n+j}{j}\binom{2j}{n-1-j}.$$

(2.4)

Combining (2.3) and (2.4), we have

Theorem 1:

$$f_n(\mathscr{B}) = \frac{1}{n+1}\sum_{j=0}^{n}\binom{n+j}{j}\binom{2j}{n-j}$$

$$g_n(\mathscr{B}) = \frac{2}{n+1}\sum_{j=0}^{n-1}\binom{n+j}{j}\binom{2j}{n-1-j}$$

$$h_n(\mathscr{B}) = \frac{1}{n+1}\sum_{j=0}^{n}\frac{j+1}{2j+1}\binom{n+j+1}{j+1}\binom{2j+1}{n-j}$$

$$S_n(\mathscr{B}) = \frac{h_n(\mathscr{B})}{\text{card } \mathscr{B}_n}$$

$$= \sum_{j=0}^{n}\frac{j+1}{2j+1}\binom{n+j+1}{j+1}\binom{2j+1}{n-j}\bigg/\binom{2n}{n}.$$

Observing the formulas in the above theorem, the question arises whether there is a combinatorial interpretation of the summands. This will be settled in a more general context in the next section.

Let us now turn to the asymptotic evaluation of the numbers appearing in Theorem 1. The common singularity $\rho = \rho(\mathscr{B})$ nearest to the origin of the generating functions $f(z)$, $g(z)$, $h(z) = 1 + f(z) + g(z)$ has been determined numerically in [6], $\rho = 0.15268...$ [compare (1.1)]. In the following, we will give the exact value of this constant, i.e., the singularity nearest to the origin of the function $u(z)$ from above.

By (2.2), ρ is a solution z of the system

$$H(z, u) = u^2(1 + u)^2 - u + z = 0$$

$$\frac{\partial H}{\partial u}(z, u) = 4u^3 + 6u^2 + 2u - 1 = 0$$

(2.5)

(Darboux's method; compare [1], [4], and [5]). Solving the second equation for u by Cardano's formula and inserting into the first equation, it turns out that

$$\rho = -y^4 + \frac{1}{2}y^2 + y - \frac{9}{16} \tag{2.6}$$

with

$$y = \frac{1}{2}\left(\sqrt[3]{1 + \sqrt{\frac{26}{27}}} + \sqrt[3]{1 - \sqrt{\frac{26}{27}}}\right).$$

Again following Darboux's method cited above, we obtain

$$[z^n]u(z) \sim c \cdot \rho^{-n} \cdot n^{-3/2}, \quad n \to \infty, \tag{2.7}$$

with

$$c = \left(\frac{\rho}{2\pi(-1 + 12y^2)}\right)^{1/2}.$$

Hence

$$f_n = [z^{n+1}]u(z) \sim \frac{c}{\rho}\rho^{-n}n^{-3/2}. \tag{2.8}$$

To determine the asymptotic behavior of g_n, we observe that

$$u(z) = u(\rho) - K(\rho - z)^{1/2} + \cdots;$$

thus,

$$u^2(z) = u^2(\rho) - 2u(\rho)K(\rho - z)^{1/2} + \cdots,$$

so that

$$g_n = [z^{n+1}]u^2(z) \sim 2u(\rho)f_n = 2\left(y - \frac{1}{2}\right)f_n. \tag{2.9}$$

Putting everything together, we arrive at

Theorem 2: With

$$y = \frac{1}{2}\left(\sqrt[3]{1 + \sqrt{\frac{26}{27}}} + \sqrt[3]{1 - \sqrt{\frac{26}{27}}}\right)$$

and

$$\rho = -y^4 + \frac{1}{2}y^2 + y - \frac{9}{16},$$

we have

$$f_n(\mathscr{B}) \sim \sqrt{\frac{1}{2\rho\pi(-1 + 12y^2)}} \cdot \rho^{-n}n^{-3/2}$$

$$\sim (0.41878180\ldots) \cdot (0.15267965\ldots)^{-n}n^{-3/2},$$

(continued)

$$g_n(\mathscr{B}) \sim 2\left(y - \frac{1}{2}\right)f_n(\mathscr{B})$$

$$\sim (0.21834433\ldots)(0.15267965\ldots)^{-n}n^{-3/2},$$

$$h_n(\mathscr{B}) \sim 2yf_n(\mathscr{B}) \sim \sqrt{\frac{2y^2}{\rho\pi(-1 + 12y^2)}} \cdot \rho^{-n}n^{-3/2}$$

$$\sim (0.63712614\ldots)(0.15267965\ldots)^{-n}n^{-3/2}.$$

In particular,

$$\frac{f_n}{g_n} \sim \frac{1}{2y - 1} = 1.917987\ldots,$$

$$S_n(\mathscr{B}) = \frac{h_n(\mathscr{B})}{\text{card } \mathscr{B}_n} \sim \sqrt{\frac{2y^2}{\rho(-1 + 12y^2)}} \cdot \left(\frac{1}{4\rho}\right)^{-n}$$

$$\sim (1.1292766\ldots)(1.6374152\ldots)^n.$$

3. t-ARY TREES

As announced in Section 2, we now determine the numbers

$$f_{n,j} = f_{n,j}(\mathscr{T}) = \sum_{T \in \mathscr{T}_n} \text{card}\{S : S \subseteq V(T);\ S \text{ a}$$

Fibonacci subset of cardinality
j *not* containing the root},

and

$$g_{n,j} = g_{n,j}(\mathscr{T}) = \sum_{T \in \mathscr{T}_n} \text{card}\{S : S \subseteq V(T);\ S \text{ a}$$

Fibonacci subset of cardinality
j containing the root},

where $\mathscr{T}_n$ denotes the family of t-ary trees of size n. Let

$$F(z, x) = \sum_{n,k} f_{n,k} z^n x^k$$

resp.

$$G(z, x) = \sum_{n,k} g_{n,k} z^n x^k$$

be the double generating functions. Since

$$\mathcal{J} = \Box + \underbrace{\overset{\displaystyle\bigwedge}{\mathcal{J}\ \ \mathcal{J} \dots \mathcal{J}}}_{t \text{ times}}, \tag{3.1}$$

it follows that

$$F = z(1 + F + G)^t$$
$$G = xz(1 + F)^t. \tag{3.2}$$

Substituting $xz = w^{t-1}$ and $V = w(1 + F)$, we have

$$G = \frac{V^t}{w}$$

and

$$F = z\left(\frac{V}{w} + \frac{V^t}{w}\right)^t = \frac{1}{xw} V^t(1 + V^{t-1})^t,$$

so that

$$V = w + wF = w + \frac{1}{x} V^t(1 + V^{t-1})^t$$

and, finally,

$$w = V\left(1 - \frac{1}{x} V^{t-1}(1 + V^{t-1})^t\right). \tag{3.3}$$

Applying LIF

$$V = \sum_k v_k(x)w^k$$

with

$$v_{k+1}(x) = \frac{1}{k+1}[y^k]\left(1 - \frac{1}{x} y^{t-1}(1 + y^{t-1})^t\right)^{-k-1}. \tag{3.4}$$

Since

$$1 + F = \frac{V}{w} = \sum_k v_{k+1}(x)w^k$$

$$= 1 + \sum_{n,k} f_{n,k} x^k z^n$$

$$= 1 + \sum_{n,k} f_{n,k}\left(\frac{1}{x}\right)^{n-k} w^{(t-1)n},$$

we have

$$\sum_{n,k} f_{n,k}\left(\frac{1}{x}\right)^{n-k} = v_{(t-1)n+1}(x)$$

$$= \frac{1}{1 + (t-1)}[y^{(t-1)n}]$$

$$\left(1 - \frac{1}{x} y^{t-1}(1 + y^{t-1})^t\right)^{-(t-1)n-1}$$

$$= \frac{1}{1 + (t - 1)n}[y^n]\left(1 - \frac{1}{x} y(1 + y)^t\right)^{-(t-1)n-1}$$

$$= \frac{1}{1 + (t - 1)n} \sum_{j=0}^{n} \binom{(t - 1)n + j}{j}\binom{tj}{n - j}\left(\frac{1}{x}\right)^t,$$

so that

$$f_{n, n-j} = \frac{1}{1 + (t - 1)n}\binom{(t - 1)n + j}{j}\binom{tj}{n - j}. \qquad (3.5)$$

In order to investigate $G = V^t/w$, we again use LIF to find that

$$V^t = \sum_{k} \tilde{v}_k (x)w^k$$

with

$$\tilde{v}_{k+1}(x) = \frac{t}{k + 1}[y^{k+1-t}]\left(1 - \frac{1}{x} y^{t-1}(1 + y^{t-1})^t\right)^{-k-1}.$$

By a similar computation,

$$g_{n, n-j} = \frac{t}{1 + (t - 1)n}\binom{(t - 1)n + j}{j}\binom{tj}{n - j - 1}. \qquad (3.6)$$

Theorem 3: The average values of the numbers of Fibonacci subsets of cardinality $n - j$ of the trees in $\mathscr{T}_n$ are given by

(a) (not containing the root)

$$\binom{(t - 1)n + j}{j}\binom{tj}{n - j}\Big/\binom{tn}{n};$$

(b) (containing the root)

$$t\binom{(t - 1)n + j}{j}\binom{tj}{n - j - 1}\Big/\binom{tn}{n};$$

(c) (in total)

$$\frac{j + 1}{tj + 1}\binom{(t - 1)n + j + 1}{j + 1}\binom{tj + 1}{n - j}\Big/\binom{tn}{n}.$$

Observe that for $t = 2$ these expressions coincide with the summands in Theorem 1, which means that the desired combinatorial interpretation may be established in this way.

Summing up over all possible values of j, we obtain the following corollary.

Corollary 1: The average Fibonacci number $S_n(\mathcal{T})$ of t-ary trees of size n is given by

$$S_n(\mathcal{T}) = \sum_{j=0}^{n} \frac{j+1}{tj+1}\binom{(t-1)n+j+1}{j+1}\binom{tj+1}{n-j}\Big/\binom{tn}{n}.$$

Before exploring the asymptotic behavior of $S_n(\mathcal{T})$ for $n \to \infty$, we want to stress the question for which value of

$$\alpha \in \left]\frac{1}{t+1}, 1\right[$$

the expression

$$h_{n,\,(1-\alpha)n}(\mathcal{T}) = \frac{\alpha n+1}{t\alpha n+1}\binom{(t-1)n+\alpha n+1}{\alpha n+1}\binom{t\alpha n+1}{n-\alpha n}$$

[compare (c) of Theorem 3] obtains its maximum for $n \to \infty$. By Stirling's approximation, we find

$$h_{n,\,(1-\alpha)n}(\mathcal{T}) \sim \frac{1}{2\pi n}\sqrt{\frac{(t-1+\alpha)^3 t}{(t-1)(1-\alpha)(\alpha(t+1)-1)^3}}\, C_{\alpha,\,t}^n,$$

with

$$C_{\alpha,\,t} = \frac{(t-1+\alpha)^{t-1+\alpha}(t\alpha)^{t\alpha}}{(t-1)^{t-1}\alpha^{\alpha}(1-\alpha)^{1-\alpha}(\alpha(t+1)-1)^{\alpha(t+1)-1}}.$$

Since $C_{\alpha,\,t}$ regulates the exponential growth, we confine our considerations to this quantity. Let α_t denote the value of α for which $C_{\alpha,\,t}$ takes its maximum. By ordinary calculus, we find that α_t must fulfil the equation

$$\frac{(\alpha_t + t - 1)\cdot t^t \cdot \alpha_t^{t-1}\cdot (1-\alpha_t)}{(\alpha_t(t+1)-1)^{t+1}} = 1. \tag{3.8}$$

For example,

$$\alpha_1 = \frac{5+\sqrt{5}}{10} = 0.7236067\ldots,$$

$$\alpha_2 = 0.7074302\ldots\,.$$

It is not difficult to see that $\alpha_\infty = \lim_{t\to\infty} \alpha_t$ exists.

Taking the logarithm in (3.8) and expanding for $t \to \infty$, it turns out that α_∞ is the (unique) solution of the equation

$$\frac{1 - \alpha_\infty}{\alpha_\infty^2} = e^{\frac{\alpha_\infty - 1}{\alpha_\infty}} \tag{3.9}$$

with $0 < \alpha_\infty < 1$, i.e.,

$$\alpha_\infty = 0.6924583\ldots . \tag{3.10}$$

By a more careful consideration, it turns out that

$$\alpha_t = \alpha_\infty + \frac{\beta}{t} + \mathcal{O}\left(\frac{1}{t^2}\right), \quad t \to \infty.$$

In a similar way as in the determination of α_∞, we find that β is given by the equation

$$\beta\left(\frac{1}{\alpha_\infty^2} + \frac{2}{\alpha_\infty} + \frac{1}{1 - \alpha_\infty}\right) = \alpha_\infty - \frac{3}{2} + \frac{1}{2\alpha_\infty^2}, \tag{3.11}$$

i.e.,

$$\beta = 0.0285962\ldots .$$

Altogether, we have proved

Theorem 4: For "large n" the maximal contribution to the average Fibonacci number $S_n(\mathscr{T})$ occurs for a cardinality $j = \gamma_t \cdot n$ of the Fibonacci subsets, where

$$\gamma_t = 1 - \alpha_t = 0.3075416\ldots - \frac{0.0285962\ldots}{t} + \mathcal{O}(1/t^2).$$

To speak in a less rigorous way, we may say that Fibonacci subsets which contain approximately 30% of the nodes of the tree constitute the maximal contribution to the Fibonacci number.

The last part of this section is devoted to the study of the asymptotic behavior of $S_n(\mathscr{T})$ for $n \to \infty$. For this reason, we introduce the generating functions

$$f(z) = F(z, 1) = \sum_n z^n \sum_k f_{n,k} ,$$

$$g(z) = G(z, 1) = \sum_n z^n \sum_k g_{n,k} . \tag{3.12}$$

From (3.2), we find that

$$f = z(1 + f + g)^t \quad \text{and} \quad g = z(1 + f)^t . \tag{3.13}$$

Substituting

$$u = z(1 + f)^{t-1},$$ (3.14)

it turns out that

$$g = u^{1+1/(t-1)} \cdot z^{-1/(t-1)}$$

and

$$f = z \left(\frac{u^{1/(t-1)}}{z^{1/(t-1)}} + \frac{u^{1+1/(t-1)}}{z^{1/(t-1)}} \right)^t$$

$$= z^{-1/(t-1)} u^{t/(t-1)} (1 + u)^t.$$

Inserting into (3.14) yields, after a few steps,

$$z = u(1 - u(1 + u)^t)^{t-1}.$$ (3.15)

In order to apply Darboux's method, we solve the system

$$H(z, u) = z - u(1 - u(1 + u)^t)^{t-1} = 0$$
$$\frac{\partial H}{\partial u}(z, u) = 0.$$ (3.16)

Let (z_t, u_t) be the pair of solutions in question. Then, after some short manipulations, the second equation (3.16) may be written as

$$tu_t(1 + tu_t)(1 + u_t)^{t-1} = 1.$$ (3.17)

From this identity, we gain the asymptotic behavior of u_t for $t \to \infty$ as follows: It is easily seen that $tu_t = \mathcal{O}(1)$. We put

$$tu_t = \delta + r_t.$$ (3.18)

Inserting and expanding, we derive

$$(1 + u_t)^{t-1} = e^\delta(1 + o(1)), \quad t \to \infty,$$

so that $\delta > 0$ is the (unique) solution of

$$\delta(1 + \delta)e^\delta = 1, \text{ i.e., } \delta = 0.4441302\ldots . $$ (3.19)

Again plugging (3.18) into (3.19), a more detailed expansion yields

$$(1 + u_t)^{t-1} = e^{\delta}\left(1 + r_t - \frac{\delta}{t} - \frac{\delta^2}{2t} + \mathcal{O}\left(\frac{1}{t}\right) + \mathcal{O}(r_t^2)\right)$$

and therefore

$$r_t = \frac{2\delta + \delta^2}{1 + 3\delta + \delta^2} \cdot \frac{\delta(\delta + 1)}{2} \cdot \frac{1}{t} + \cdots, \quad (t \to \infty),$$

so that

$$u_t = \frac{\delta}{t} + \frac{\varepsilon}{t^2} + \mathcal{O}\left(\frac{1}{t^3}\right) \tag{3.20}$$

with

$$\varepsilon = \frac{\delta^2(\delta + 1)(\delta + 2)}{2((\delta + 1)(\delta + 2) - 1)} = 0.1376138\ldots .$$

Turning now to z_t, (3.15) combined with (3.17) yields

$$z_t = u_t\left(1 - \frac{1 + u_t}{t(1 + tu_t)}\right)^{t-1}$$

$$= u_t\left(\frac{t - 1}{t}\right)^{t-1}\left(1 + \frac{u_t}{1 + tu_t}\right)^{t-1}$$

$$= tu_t \cdot q_t\left(1 + \frac{u_t}{1 + tu_t}\right)^{t-1}, \tag{3.21}$$

where

$$q_t = \frac{1}{t}\left(\frac{t - 1}{t}\right)^{t-1} \tag{3.22}$$

is the unique singularity nearest to the origin of the generating function

$$y(z) = \sum_{n \geq 0} \frac{1}{1 + (t - 1)n}\binom{tn}{n}z^n$$

of the numbers of trees in $\mathcal{I}_n$.

By Darboux's theorem, it follows that $S_n(\mathcal{I})$ behaves like

$$S_n(\mathcal{I}) \sim A_t\left(\frac{q_t}{z_t}\right)^n, \quad n \to \infty, \tag{3.23}$$

where A_t is a constant that will not be determined explicitly here, for shortness. The ratio q_t/z_t [i.e., the order of growth of $S_n(\mathcal{I})$] behaves for $t \to \infty$, by (3.20) and (3.21), as

$$\frac{q_t}{z_t} = (\delta + 1)e^{\delta^2/(\delta+1)}\left(1 + \frac{1}{t}\left(\frac{\delta}{\delta + 1} + \frac{\delta^2}{2(\delta + 1)^2} - \frac{\varepsilon}{(\delta + 1)^2} - \frac{\varepsilon}{\delta}\right) + \cdots\right).$$

Evaluating the appearing constants numerically, we get

Theorem 5: With A_t a constant, we have

$$S_n(\mathcal{T}) \sim A_t\left(1.655487\ldots - \frac{0.0489690\ldots}{t} + \mathcal{O}\!\left(\frac{1}{t^2}\right)\right)$$

for $n \to \infty$.

4. PLANTED PLANE TREES

The family $\mathcal{P}$ of planted plane trees is defined by the following symbolic equation:

$$\mathcal{P} = \circ + \begin{array}{c} \circ \\ | \\ \mathcal{P} \end{array} + \begin{array}{c} \circ \\ \wedge \\ \mathcal{P}\ \mathcal{P} \end{array} + \begin{array}{c} \circ \\ \wedge \\ \mathcal{P}\ \mathcal{P}\ \mathcal{P} \end{array} + \cdots \ . \qquad (4.1)$$

Let us denote by $f_{n,j} = f_{n,j}(\mathcal{P})$, $g_{n,j} = g_{n,j}(\mathcal{P})$ the numbers of Fibonacci subsets of cardinality j of the trees of size n in $\mathcal{P}$ (not containing resp. containing the root) and by $F(z, x)$ resp. $G(z, x)$ the double generating functions. From (4.1), we obtain

$$F = \frac{z}{1 - F - G}, \quad G = \frac{zx}{1 - F}. \qquad (4.2)$$

From this

$$z = \frac{F(1 - F)^2}{1 + F(x - 1)} \qquad (4.3)$$

Applying LIF as in the previous section, we obtain

$$f_{n,j} = \frac{1}{n}\binom{n}{j}\binom{2n - 2}{n - j - 1}$$

$$g_{n,j} = \frac{1}{n - 1}\binom{n - 1}{j - 1}\binom{2n - 2}{n - j - 1}. \qquad (4.4)$$

Theorem 6: The average numbers of Fibonacci subsets of cardinality j of planted plane trees of size n are given by:

(a) (not containing the root)

$$\binom{n}{j}\binom{2n - 2}{n - j - 1}\bigg/\binom{2n - 2}{n - 1};$$

(b)   (containing the root)

$$\frac{n}{n-1}\binom{n-1}{j-1}\binom{2n-2}{n-j-1}\Big/\binom{2n-2}{n-1};$$

(c)   (in total)

$$2\binom{n}{j}\binom{2n-3}{n-j-1}\Big/\binom{2n-2}{n-1}.$$

Applying Vandermonde's convolution, we obtain

Corollary 2: The average numbers of Fibonacci subsets of planted plane trees of size n are given by:

(a)   (not containing the root)

$$a_n := \binom{3n-2}{n-1}\Big/\binom{2n-2}{n-1};$$

(b)   (containing the root)

$$b_n := \frac{n}{n-1}\binom{3n-3}{n-2}\Big/\binom{2n-2}{n-1};$$

(c)   (in total)

$$2\binom{3n-3}{n-1}\Big/\binom{2n-2}{n-1} \sim \sqrt{3}\cdot\left(\frac{27}{16}\right)^{n-1}, \quad (n \to \infty);$$

(d) $\dfrac{a_n}{b_n} = 3 - \dfrac{2}{n}$.

The second-order moments of all random variables in question are not much harder to obtain than the expected values. To give an example, we determine the second-order moment in the case of planted plane trees.

Let $f(T)$ resp. $g(T)$ denote the number of Fibonacci subsets of the tree T not containing resp. containing the root and

$$A(z) = \sum_n z^n \sum_{T \in \mathscr{P}_n} (f(T) + g(T))^2;$$

$$B(z) = \sum_n z^n \sum_{T \in \mathscr{P}_n} f^2(T);$$

(4.4)

(continued)

$$C(z) = \sum_n z^n \sum_{T \in \mathscr{P}_n} f(T)g(T);$$

$$(4.4)$$

$$D(z) = \sum_n z^n \sum_{T \in \mathscr{P}_n} g^2(T).$$

So we have

$$A = B + 2C + D,$$ $$(4.5)$$

and, by (4.1),

$$B = z/(1 - A),$$
$$C = z/(1 - B - C),$$ $$(4.6)$$
$$D = z/(1 - B).$$

From (4.6), it follows that

$$B = \cfrac{z}{1 - B - 2C - \cfrac{z}{1 - B}},$$

or

$$z = B(1 - B)(1 - B - 2C),$$

whence

$$2C = 1 - B - \frac{z}{B(1 - B)}.$$

Inserting this into

$$4z = 2C(2(1 - B) - 2C),$$

we derive

$$4z = (1 - B)^2 - \frac{z^2}{B^2(1 - B)^2},$$

or

$$z = \frac{B}{\varphi(B)} \quad \text{with } \varphi(B) = (1 - B)^{-2}(-2B + \sqrt{1 + 4B^2})^{-1}.$$
$$(4.7)$$

Applying LIF,

$$[z^n]B = \frac{1}{n}[z^{n-1}](1 - z)^{-2n}(2z + \sqrt{1 + 4z^2})^n.$$

Substituting $z = u/(1 - u^2)$, it follows by formal residue calculation that

$$[z^n]B = \frac{1}{n}[u^{n-1}]\frac{(1+u^2)(1+u)^{4n-2}(1-u)^{2n-2}}{(1-u-u^2)^{2n}}, \qquad (4.8)$$

whence

$$[z^n]B = \frac{1}{n}\sum_{i+\ell+j=n-1}\binom{2n+i-1}{2n-1}\binom{n-j}{n-\ell-j}$$

$$\left[\binom{n+\ell+j-1}{j} + \binom{n+\ell+j-1}{j-2}\right]. \qquad (4.9)$$

Similarly,

$$[z^n] = \frac{1}{n}[u^{n-1}]\frac{(1-3u)(1+u)^{4n-3}(1-u)^{2n-2}}{(1-u-u^2)^{2n}}, \qquad (4.10)$$

$$[z^n] = \frac{1}{n}[u^{n-1}]$$

$$\frac{(1+u^2)(1-2u-2u^2-2u^3-u^4)(1+u)^{4n-6}(1-u)^{2n-2}}{(1-u-u^2)^{2n}}$$

$$\qquad (4.11)$$

(and $A = B + 2C + D!$).

To perform the asymptotics of $[z^n]A$, we again use Darboux's method. Starting from (4.7), the method already described in the previous sections leads to the numerical value

$$q = 0.08738321...$$

for the singularity q of B (and also C, D, A) nearest to the origin. By local expansions of the generating functions about the singularity q, a tedious computation leads to (compare [5])

Theorem 7:

$$[z^n]A \sim \frac{1.755746...}{2\sqrt{\pi}}\, q^{-n+1/2}\, n^{-3/2}$$

and the second-order moment of the number of Fibonacci sub-sets is asymptotically given by

$$\frac{[z^n]A}{\text{card } \mathscr{P}_n} \sim (1.038020...)(2.860961...)^n.$$

REFERENCES

[1] Bender, E. A. 'Asymptotic Methods in Enumeration.'
     SIAM Review 16 (1974):485–515.
[2] Bruijn, N. G. de. *Asymptotic Methods in Analysis.*
     Amsterdam: North Holland, 1958.
[3] Goulden, I. P., and Jackson, D. M. *Combinatorial Enu-*
     meration. New York: John Wiley & Sons, 1983.
[4] Greene, D. H., and Knuth, D. E. *Mathematics for the*
     Analysis of Algorithms. Basel: Birkhäuser Verlag,
     1981.
[5] Harary, F., and Palmer, E. M. *Graphical Enumeration.*
     New York and London: Academic Press, 1973.
[6] Kirschenhofer, P.; Prodinger, H.; and Tichy, R. F.
     'Fibonacci Numbers of Graphs II.' *The Fibonacci Quar-*
     terly 21, no. 3 (1983):219–229.
[7] Knuth, D. E. *The Art of Computer Programming I, Funda-*
     mental Algorithms. New York: Addison Wesley, 1968.
[8] Prodinger, H., and Tichy, R. F. 'Fibonacci Numbers of
     Graphs.' *The Fibonacci Quarterly* 20, no. 1 (1982): 16–
     21.

Péter Kiss

A DISTRIBUTION PROPERTY OF SECOND-ORDER LINEAR RECURRENCES

INTRODUCTION AND RESULTS

A second-order linear recurrence $R = (R_n)$, $n = 0, 1, 2, \ldots$, is defined by integers A_1, A_2, R_0, R_1 and by the recursion

$$R_n = A_1 \cdot R_{n-1} + A_2 \cdot R_{n-2}$$

for $n > 1$. We suppose that $A_1 A_2 \neq 0$ and not both R_0 and R_1 are zero. If α and β denote the roots of the characteristic polynomial $x^2 - A_1 x - A_2$ of the sequence R and α/β is not a root of unity, then R is called a nondegenerate sequence. For nondegenerate second-order linear recurrences, we have

$$R_n = a \cdot \alpha^n + b \cdot \beta^n \tag{1}$$

for $n = 0, 1, 2, \ldots$, where

$$a = \frac{R_1 - R_0 \beta}{\alpha - \beta} \quad \text{and} \quad b = \frac{R_1 - R_0 \alpha}{\alpha - \beta}. \tag{2}$$

If $A_1 = A_2 = R_1 = 1$ and $R_0 = 0$, then the sequence R is called a Fibonacci sequence, and we shall denote it by $F = (F_n)$.

The distribution properties of linear recurrences were investigated by several authors. For example, we know that the sequence $(\log F_n)$ is uniformly distributed mod 1 (see R. L. Duncan [2] and L. Kuipers [5]). This result was extended by L. Kuipers and J.-S. Shiue [7] for linear recurrences of order greater than two. Some results gave conditions for integers m for which the sequence R is uniformly distributed mod m (e.g., see P. Bundschuh and J.-S. Shiue [1] or H. Niederreiter [9]. For general recurrences in a finite field, H. Niederreiter and J.-S. Shiue [10, 11] obtained results. Some other results can be found by the references of the papers mentioned above.

121

A. N. Philippou et al. (eds.), Fibonacci Numbers and Their Applications, 121–130.
© 1986 by D. Reidel Publishing Company.

Another type of results were obtained concerning the numbers of the form R_i/R_j, where R_i and R_j are terms of a nondegenerate second-order linear recurrence R supposing that $R_j \neq 0$. F. Mátyás [8] proved that the fractional parts of these numbers are not dense everywhere on the unit interval $[0, 1)$ if $A_1^2 + 4A_2 > 0$, that is, if α and β are real numbers. In [4] we showed that in the case $A_1^2 + 4A_2 < 0$ the fractional parts of the terms of the sequence (R_{n+1}/R_n) are everywhere dense on the unit interval $[0, 1)$ but the sequence is not uniformly distributed mod 1.

The purpose of this paper is to study another distribution property of second-order linear recurrences.

Let $A = (a_n)$, $n = 1, 2, \ldots$, be an infinite sequence of integers and let N, m, and j be integers with $m \geq 2$. Denote by $A(N, j, m)$ the number of terms a_n satisfying the conditions $1 \leq n \leq N$ and $a_n \equiv j \pmod{m}$. If the limit

$$\lim_{N \to \infty} \frac{A(N, j, m)}{N} = \frac{1}{m}$$

exists for all j with $1 \leq j \leq m$, then the sequence is called uniformly distributed modulo m. The sequence A is said to be relatively measurable modulo m if the limit

$$\lim_{N \to \infty} \frac{A(N, j, m)}{N} = \gamma_j \tag{3}$$

exists for every $j = 1, 2, \ldots, m$ (see [6], p. 315). Naturally, a sequence is uniformly distributed mod m if it is relatively measurable with $\gamma_j = 1/m$ for every $j = 1, 2, \ldots, m$.

In the following, we shall investigate the integer part sequence of the numbers R_{n+1}/R_n. If $A_1^2 + 4A_2 > 0$, i.e., α and β are real numbers, then

$$\lim_{n \to \infty} (|R_{n+1}/R_n|) = \max(|\alpha|, |\beta|)$$

and so the properties of this sequence can be easily characterized. In the case $A_1^2 + 4A_2 < 0$, the following result is true.

Theorem: Let $R = (R_n)$ be a nondegenerate second-order linear recurrence with condition $A_1^2 + 4A_2 < 0$ and let $G = (G_n)$, $n = 1, 2, \ldots$, be an infinite sequence of integers defined by

$$G_n = [R_{n+1}/R_n],$$

where $[x]$ denotes the integer part function, and we assume that $R_n \neq 0$. Then the sequence G is relatively measurable mod m for every integer $m \geqslant 2$. The sequence G is uniformly distributed mod m if and only if $m = 2$ and A_1 is an even integer.

For the proof ot the Theorem, we need a result which corresponds to a result of S. Uchiyama [12]. He proved that a sequence $A = (a_n)$ is uniformly distributed mod m ($m \geqslant 2$) if and only if

$$\lim_{N \to \infty} \frac{1}{N} \sum_{n=1}^{N} e\left(a_n \cdot \frac{h}{m}\right) = 0$$

for all $h = 1, 2, \ldots, m - 1$, where

$$e(x) = \exp(2\pi i x).$$

Following Uchiyama's argument, we shall prove the proposition below.

Proposition: Let $A = (a_n)$ be an infinite sequence of integers and let $m \geqslant 2$ be an integer. Then A is relatively measurable mod m if and only if the limit

$$\lim_{N \to \infty} \frac{1}{N} \sum_{n=1}^{N} e\left(a_n \cdot \frac{h}{m}\right) = \delta_h$$

exists for every integer h.

We note that our Proposition implies two consequences.

Corollary 1: Let A be a relatively measurable sequence mod m. If γ_j and δ_h are the numbers defined in (3) and in the Proposition, respectively, then

$$\delta_h = \sum_{j=1}^{m} \gamma_j \cdot e\left(j \cdot \frac{h}{m}\right)$$

and

$$\gamma_j = \frac{1}{m} \sum_{h=1}^{m} \delta_h \cdot e\left(h \cdot \frac{m-j}{m}\right)$$

for every integer h and $j = 1, 2, \ldots, m$.

Corollary 2: If A is a relatively measurable sequence mod m, then

$$\sum_{h=1}^{m} \delta_h = m \cdot \gamma_m,$$

where γ_m and δ_h are the numbers defined in (3) and in the Proposition, respectively.

PROOFS

First we prove the Proposition and its corollaries.

Proof of the Proposition: Let $A = (a_n)$ be an infinite sequence of integers and let $m \geqslant 2$ be an integer. First we suppose that

$$\lim_{N \to \infty} \frac{A(N, j, m)}{N} = \gamma_j$$

exists for every $j = 1, 2, \ldots, m$. Then, for an integer h, we have

$$\frac{1}{N} \sum_{n=1}^{N} e\left(a_n \cdot \frac{h}{m}\right) = \frac{1}{N} \sum_{j=1}^{m} A(N, j, m) \cdot e\left(j \cdot \frac{h}{m}\right)$$

$$= \sum_{j=1}^{m} (\gamma_j + o(1)) \cdot e\left(j \cdot \frac{h}{m}\right)$$

$$= \sum_{j=1}^{m} \gamma_j \cdot e\left(j \cdot \frac{h}{m}\right) + o(1).$$

This implies that the limit

$$\lim_{N \to \infty} \frac{1}{N} \sum_{n=1}^{N} e\left(a_n \cdot \frac{h}{m}\right) = \sum_{j=1}^{m} \gamma_j \cdot e\left(j \cdot \frac{h}{m}\right) = \delta_h \qquad (4)$$

exists for every integer h.
Conversely, let us suppose that

$$\lim_{N \to \infty} \frac{1}{N} \sum_{n=1}^{N} e\left(a_n \cdot \frac{h}{m}\right) = \delta_h \qquad (5)$$

exists for all integers h. Naturally, $\delta_m = 1$, and we can assume that $1 \leqslant h \leqslant m$ since $\delta_{mk+h} = \delta_h$ for any integer k. By (5), we have

$$\sum_{n=1}^{N} e\left(a_n \cdot \frac{h}{m}\right) = \sum_{j=1}^{m} A(N, j, m) \cdot e\left(j \cdot \frac{h}{m}\right)$$

$$= \delta_h \cdot N + o(N). \tag{6}$$

Furthermore, we know that

$$\sum_{h=1}^{m} e\left(t \cdot \frac{h}{m}\right) = 0 \tag{7}$$

if $t \not\equiv 0 \pmod{m}$ and

$$\sum_{h=1}^{m} e\left(t \cdot \frac{h}{m}\right) = m \tag{8}$$

if $t \equiv 0 \pmod{m}$. Using (6), (7), and (8) for an integer j with $1 \leqslant j \leqslant m$, we get

$$A(N, j, m) = \frac{1}{m} \sum_{k=1}^{m} A(N, k, m) \sum_{h=1}^{m} e\left((k - j)\frac{h}{m}\right)$$

$$= \frac{1}{m} \sum_{h=1}^{m} e\left(-j \cdot \frac{h}{m}\right) \sum_{k=1}^{m} A(N, k, m) \cdot e\left(k \cdot \frac{h}{m}\right)$$

$$= \frac{1}{m} \sum_{h=1}^{m} e\left(-j \cdot \frac{h}{m}\right)(\delta_h N + o(N))$$

$$= N\left[\frac{1}{m} \sum_{h=1}^{m} \delta_h \cdot e\left(-j \cdot \frac{h}{m}\right) + o(1)\right].$$

From this, it follows that limit (3) exists and

$$\gamma_j = \frac{1}{m} \sum_{h=1}^{m} \delta_h \cdot e\left(-j \cdot \frac{h}{m}\right); \tag{9}$$

thus, the sequence A is really relatively measurable mod m.

Proof of Corollary 1: The statements follow from (4) and (9) using the fact that

$$e\left(-j \cdot \frac{h}{m}\right) = e\left(h \cdot \frac{m - j}{m}\right).$$

Proof of Corollary 2: In case $j = m$, by (9), we have

$$\gamma_m = \frac{1}{m} \sum_{h=1}^{m} \delta_h$$

and so

$$\sum_{h=1}^{m} \delta_h = m \cdot \gamma_m.$$

Now, using the proposition, we can prove the Theorem.

Proof of the Theorem: Let $m \geqslant 2$ be an integer and let G be a sequence satisfying the conditions of the Theorem. By the conditions, the roots of the characteristic polynomial $x^2 - A_1 x - A_2$ are nonreal complex numbers and one of them is the conjugate of the other. Thus, α and β can be written in the forms

$$\alpha = re^{i\theta} \quad \text{and} \quad \beta = re^{-i\theta},$$

where r and θ are nonzero real numbers and $r > 0$. Similarly, by (2), we have

$$a = r_1 e^{i\omega} \quad \text{and} \quad b = \overline{a} = r_1 e^{-i\omega},$$

with some real r_1 and ω. We note that $a \neq 0$ and $b \neq 0$, since otherwise by (2), both R and R_1 would be zero or α and β would be real numbers. We can assume that $R_n \neq 0$ for any $n \geqslant 0$. Namely, we proved in [3] that there is a number n_0 depending only on A_1, A_2, R_0, and R_1 such that $R_n \neq 0$ for $n \geqslant n_0$; furthermore, the terms of the sequence R for large indices are uniquely determined by any two consecutive terms, e.g., R_{n_0} and R_{n_0+1}, as initial terms. Using the forms of α, β, a, and b, by (1) we get

$$\frac{R_{n+1}}{R_n} = \frac{r_1 r^{n+1} e^{i(\omega + (n+1)\theta)} + r_1 r^{n+1} e^{-i(\omega + (n+1)\theta)}}{r_1 r^n e^{i(\omega + n\theta)} + r_1 r^n e^{-i(\omega + n\theta)}}$$

$$= r \cdot \frac{\cos(\omega + (n+1)\theta)}{\cos(\omega + n\theta)}$$

$$= r(\cos \theta - \sin \theta \cdot tg(n\theta + \omega))$$

$$= c + d \cdot tg(n\theta + \omega),$$

where
$$c = r \cos \theta$$
and
$$d = -r \sin \theta$$

are nonzero real numbers.
Let
$$D(N, h, m) = \frac{1}{N} \sum_{n=1}^{N} e\left(h \cdot \frac{[c + d \cdot tg(n\theta + \omega)]}{m}\right),$$

where $[x]$ denotes the integer part of the real number x. By the Proposition, we have to prove that the limit

$$\lim_{N \to \infty} D(N, h, m) = \delta_h$$

exists for all integers h with $1 \leqslant h \leqslant m$.

First we suppose that $d > 0$. If k is an integer and

$$G_n = [c + d \cdot tg(n\theta + \omega)] = k,$$

then

$$k \leqslant c + d \cdot tg(n\theta + \omega) < k + 1,$$

which can be written in the form

$$\frac{k - c}{d} \leqslant tg(n\theta + \omega) = tg\left(\pi\left(n\frac{\theta}{\pi} + \frac{\omega}{\pi}\right)\right) < \frac{k + 1 - c}{d}.$$
$$(10)$$

Let g_n be a real number for which $-1/2 \leqslant g_n < 1/2$ and

$$n\frac{\theta}{\pi} + \frac{\omega}{\pi} = g_n + q_n$$

with some integer q_n. Then (10) is equivalent to the inequalities

$$\frac{1}{\pi} \text{Arc } tg \, \frac{k - c}{d} \leqslant g_n < \frac{1}{\pi} \text{Arc } tg \, \frac{k + 1 - c}{d}.$$
$$(11)$$

Let $B(N, k)$ denote the number of integers n for which $1 \leqslant n \leqslant N$ and (11) holds. Using this notation, we can write

$$D(N, h, m) = \sum_{k=-\infty}^{+\infty} \frac{B(N, k)}{N} \cdot e\left(h \cdot \frac{k}{m}\right)$$

$$= \sum_{j=1}^{m} \left(e\left(h \cdot \frac{j}{m}\right) \sum_{n=-\infty}^{+\infty} \frac{B(N, j + nm)}{N}\right).$$
$$(12)$$

We have assumed that α/β is not a root of unity and so θ/π is an irrational number. Therefore, the sequence

$$\left(n\frac{\theta}{\pi} + \frac{\omega}{\pi}\right), \quad n = 1, 2, \ldots,$$

is uniformly distributed mod 1 and it is uniformly distributed on the interval $[-1/2, 1/2)$, too. This implies that

$$\frac{B(N, k)}{N} = \frac{1}{\pi}\left(\text{Arc } tg \frac{k+1-c}{d} - \text{Arc } tg \frac{k-c}{d}\right) + o(1)$$

for any integer k. Thus, by (12) we have

$$\lim_{N \to \infty} D(N, h, m) = \frac{1}{\pi}\sum_{j=1}^{m} e\left(h \cdot \frac{j}{m}\right)S_j, \qquad (13)$$

where

$$S_j = \sum_{n=-\infty}^{+\infty} \left(\text{Arc } tg \frac{j+nm+1-c}{d} - \text{Arc } tg \frac{j+nm-c}{d}\right).$$
$$(14)$$

But the sum S_j is convergent for every $j = 1, 2, \ldots, m$, so the limit (13) exists and the sequence G is really relatively measurable mod m.

We shall show that G is not uniformly distributed mod m if $m > 2$. By Uchiyama's criterion, it is enough to prove that, in the case $m > 2$,

$$\lim_{N \to \infty} D(N, h, m) = \delta_h \neq 0$$

for some integer h for which $h \not\equiv 0 \pmod{m}$.

By the definition of α, we have

$$\alpha = \frac{A_1}{2} + \frac{i}{2}\sqrt{-A_1^2 - 4A_2}$$

and so

$$c = |\alpha| \cos \theta = \text{Re}(\alpha) = \frac{A_1}{2}.$$

Thus, for some integer t,

$$c = t \quad \text{or} \quad c = t + \frac{1}{2},$$

according as A_1 is even or odd. In both cases, by (13) and (14) it follows that $\delta_1 \neq 0$, using that

$$\text{Arc } tg \; x = -\text{Arc } tg(-x)$$

and that $S_{j+t} = S_{-j+t-1}$ in case $c = t$ and $S_{j+t} = S_{-j+t}$ in case $c = t + 1/2$ for every integer j.

If $m = 2$ and A_1 is an odd integer, then $c = t + 1/2$. In this case, by (14) one can easily see that $S_1 \neq S_2$, and from this,

$$\lim_{N \to \infty} D(N, 1, 2) \neq 0$$

follows. Thus, in this case also, the sequence is not uniformly distributed mod m.

If $m = 2$ and A_1 is even, that is, c is an integer, then $S_1 = S_2 = \pi/2$ and so, by $e(j/2) = \pm 1$, using (13), $\delta_1 = 0$ follows. Thus, in this case, the sequence   is uniformly distributed mod m.

The case $d < 0$ can be traced back to the case $d > 0$, which completes the proof of the Theorem.

ACKNOWLEDGMENT

The author would like to express his gratitude to Professor P. Bundschuh for his valuable advice, which led to the writing of this paper.

REFERENCES

[1] Bundschuh, P., and Shiue, J.-S. 'Solution of a Problem on the Uniform Distribution of Integers.' *Atti Accad. Naz. Lincei, Rend, Cl. Sci. Fis. Mat. Nat.* **55** (1973): 172–177.

[2] Duncan, R. L. 'Application of Uniform Distribution to the Fibonacci Numbers.' *The Fibonacci Quarterly* **5**, no. 2 (1967):137–140.

[3] Kiss, P. 'Zero Terms in Second Order Linear Recurrences.' *Math. Sem. Notes* (Kobe Univ) 7 (1979):145–152.

[4] Kiss, P. 'Distribution of the Ratios of the Terms of Binary Recurrences. To appear.

[5] Kuipers, L. 'Remark on a Paper by R. L. Duncan Concerning the Uniform Distribution mod 1 of the Sequence of the Logarithms of the Fibonacci Numbers.' *The Fibonacci Quarterly* **7**, no. 5 (1969):465–466, 473.

[6] Kuipers, L., and Niederreiter, H. *Uniform Distribution of Sequences.* New York: John Wiley & Sons, 1974.

[7] Kuipers, L., and Shiue, J.-S.   'Remark on a Paper by
     Duncan and Brown on the Sequence of Logarithms of Cer-
     tain Recursive sequences.' *The Fibonacci Quarterly* **11**,
     no. 4 (1973):292-294.
[8] Matyas, F.   'On the Quotients of the Elements of Linear
     Recursive Sequences of Second Order' (in Hungarian).
     Mat. Lapok **27** (1976-1979):379-389.
[9] Niederreiter, H.   'Distribution of Fibonacci Numbers
     mod 5^k.' *The Fibonacci Quarterly* **10**, no. 4 (1972):
     373-374.
[10] Niederreiter, H., and Shiue, J.-S.   'Equidistribution
     of Linear Recurring Sequences in Finite Fields I.'
     Math. **80** (1977):397-405.
[11] Niederreiter, H., and Shiue, J.-S.   'Equidistribution
     of Linear Recurring Sequences in Finite Fields II.'
     Acta Arith. **38** (1980):197-207.
[12] Uchiyama, S.   'On the Uniform Distribution of Sequences
     of Integers.' *Proc. Japan Acad.* **37** (1961):605-609.

Péter Kiss, Bui Minh Phong, and Erik Lieuwens

ON LUCAS PSEUDOPRIMES WHICH ARE
PRODUCTS OF s PRIMES

1. INTRODUCTION AND RESULTS

Let P and Q be nonzero integers such that $(P, Q) = 1$ and $D = P^2 - 4Q \neq 0$. A Lucas sequence (U_n) is defined by initial terms $U_0 = 0$, $U_1 = 1$, and by the recursion

$$U_n = P \cdot U_{n-1} - Q \cdot U_{n-2}$$

for $n > 1$. Its associate sequence (V_n) is given by $V_0 = 2$, $V_1 = P$, and

$$V_n = P \cdot V_{n-1} - Q \cdot V_{n-2}$$

for $n > 1$. In the following we assume that the sequences are nondegenerate, i.e., $U_n \neq 0$ and $V_n \neq 0$ if $n \geqslant 1$.

For odd primes n with $(n, QD) = 1$, as is well known, we have

$$U_{n-(D/n)} \equiv 0 \pmod{n}, \tag{1}$$

$$U_n \equiv (D/n) \pmod{n}, \tag{2}$$

and

$$V_n \equiv V_1 \pmod{n}, \tag{3}$$

where (D/n) is the Jacobi symbol. If n is composite, but (1) still holds, then we call n a Lucas pseudoprime with parameters P and Q. We say n is a Euler–Lucas pseudoprime with parameters P and Q if n is an odd composite integer $(n, QD) = 1$ and

$$U_{\frac{n-(D/n)}{2}} \equiv 0 \pmod{n} \text{ if } (Q/n) = 1$$

or

$$V_{\frac{n-(D/n)}{2}} \equiv 0 \pmod{n} \text{ if } (Q/n) = -1.$$

Lucas and Euler–Lucas pseudoprimes are generalizations of pseudoprimes and Euler pseudoprimes to base an integer b (> 1), respectively; namely, a composite n is called a

131

A. N. Philippou et al. (eds.), Fibonacci Numbers and Their Applications, 131–139.
© 1986 by D. Reidel Publishing Company.

pseudoprime to base b if $(n, b) = 1$ and

$$b^{n-1} \equiv 1 \pmod{n},$$

and we say n is a Euler pseudoprime to base b if

$$b^{(n-1)/2} \equiv (b/n) \pmod{n}.$$

In case $b = 2$, we only say n is a pseudoprime or a Euler pseudoprime.

The properties of pseudoprimes have been studied intensively because they can be used for primality tests (e.g., see [1]). We list some results that are connected with ours. C. Pomerance, J. L. Selfridge, and S. S. Wagstaff, Jr. [8] proved that there exist infinitely many Euler pseudoprimes and there are Euler pseudoprimes with arbitrarily many prime divisors. E. A. Parberry [7] has shown that there are infinitely many Euler-Lucas pseudoprimes with parameters 1 and -1 (i.e., U_n is the Fibonacci sequence). Some results of A. Rotkiewicz [9] imply that in case $D > 0$ every arithmetic progression $ax + b$ ($x = 0, 1, 2, \ldots$), where $(a, b) = 1$, contains an infinite number of Euler-Lucas pseudoprimes with parameters P and Q; this result is an improvement of a result of R. Baillie and S. S. Wagstaff, Jr. [1]. On the other hand, it is known that for all integers b, $s \geqslant 2$ there exist infinitely many pseudoprimes to base b which are products of exactly s distinct primes (see P. Erdös [3] and E. Lieuwens [6]). Lieuwens stated that this result can be generalized for Lucas pseudoprimes, too.

The purpose of this paper is to give an extension of the results mentioned above. We shall prove the following.

Theorem 1: Let (U_n) be a nondegenerate Lucas sequence and let s, k ($\geqslant 2$) be integers. Then there exist infinitely many Euler-Lucas pseudoprimes with parameters P and Q which are products of exactly s distinct primes. If $D = P^2 - 4Q > 0$, then the primes can be chosen from the arithmetic progression $kx + 1$ ($x = 1, 2, \ldots$).

This theorem immediately implies the following consequence.

Corollary: For all integers b, k, $s \geqslant 2$ there exist infinitely many Euler pseudoprimes to base b which have the form $kx + 1$ and are products of s distinct primes.

The proof of Theorem 1 implies another consequence. A. Rotkiewicz [10] proved that if $Q = 1$ or $Q = -1$ then there are infinitely many pairs p, q of distinct primes such that the numbers $n = pq$ satisfy congruences (1), (2), and (3) simultaneously; furthermore, in every arithmetic progression $ax + b$, where $(a, b) = 1$, there exist infinitely many composite n for which (1), (2), and (3) hold simultaneously. We show a similar result.

Theorem 2: Let (U_n) be a nondegenerate Lucas sequence with $Q = 1$ or $Q = -1$. Then for all integers k, $s \geqslant 2$ there exist infinitely many Euler pseudoprimes which satisfy congruences (1), (2), and (3) simultaneously and which are products of s distinct primes of the form $kx + 1$.

2. ELEMENTARY PROPERTIES OF SEQUENCES (U_n)

First we recall some known results on Lucas sequences and prove two lemmas that will be used in the proofs of our theorems.

Let (U_n) be a nondegenerate Lucas sequence defined by parameters P, Q, and $D = P^2 - 4Q$. It is known that for any nonzero integer n with $(n, Q) = 1$ there are terms in (U_n) divisible by n. The least positive integer r, for which $n|U_r$, is called the rank of apparition of n in the sequence, and we shall denote it by $r(n)$. If $n = p$ is a prime, then p is called a primitive prime divisor of $U_{r(p)}$ or, more exactly, p is a primitive prime divisor of U_n if $p|U_n$ but $p \nmid QDU_m$ for $0 < m < n$. It is known that there is an absolute constant n_0 such that U_n has at least one primitive prime divisor for every $n \geqslant n_0$ (see A. Schinzel [11] or C. L. Stewart [12]).

Let m and n be positive integers with $(mn, QD) = 1$ and let p be a prime for which $(p, QD) = 1$. Using the notations defined above, we have

(i) $n|U_m$ if and only if $r(n)|m$,

(ii) $r(p)|(p - (D/p))$,

(iii) $r(p) \left| \dfrac{p - (D/p)}{2} \right.$ if and only if $(Q/p) = 1$,

(iv) $r([m, n]) = [r(m), r(n)]$,

where $[x, y]$ denotes the least common multiple of x and y.
(For these properties of Lucas sequences, we refer to D. H.
Lehmer [5] and H. J. A. Duparc [2]).

We shall prove some other properties of Lucas
sequences.

Lemma 1: Let p be a prime number with $(p, 2QD) = 1$.   If
$8QD \,|\, r(p)$, then

$$(Q/p) = 1 \quad \text{and} \quad r(p) \left| \frac{p - (D/p)}{2} \right. ;$$

furthermore, $(D/p) = 1$ if $D > 0$.

Proof of Lemma 1: Since $QD \neq 0$, we can write $D = \pm 2^u D'$ and
$Q = \pm 2^v Q'$, where $u, v \geqslant 0$ and D' and Q' are positive odd
integers.   If $8QD \,|\, r(p)$ for a prime p with $(p, 2QD) = 1$, then
by (ii) p is of the form

$$p = 8QDy + (D/p),$$

where y is an integer, and so $(2^u/p) = (2^v/p) = 1$, $(-1/p) = (D/p)$, and $(p/q) = ((D/p)/q)$ for any odd divisor q of QD.

First, let $D > 0$.   In this case

$$
\begin{aligned}
(D/p) &= (2^u/p) \cdot (D'/p) \\
&= (-1)^{\frac{D'-1}{2} \cdot \frac{p-1}{2}} \cdot (p/D') \\
&= (-1/p)^{\frac{D'-1}{2}} \cdot ((D/p)/D') \\
&= (D/p)^{\frac{D'-1}{2}} \cdot (D/p)^{\frac{D'-1}{2}} = 1.
\end{aligned}
$$

Thus, p has the form $p = 8k + 1$, where $Q' \,|\, k$, and so

$$(Q/p) = (Q'/p) = (p/Q') = 1. \tag{4}$$

If $D < 0$, then $Q > 0$ and, similarly to the above case,
we have

$$(Q/p) = (2^v/p) \cdot (Q'/p) = (Q'/p)$$

$$= (-1)^{\frac{Q'-1}{2} \cdot \frac{p-1}{2}} \cdot (p/Q')$$

$$= (-1/p)^{\frac{Q'-1}{2}} \cdot ((D/p)/Q')$$

$$= (D/p)^{\frac{Q'-1}{2}} \cdot (D/p)^{\frac{Q'-1}{2}} = 1. \qquad (5)$$

Thus, by (4) and (5), $(Q/p) = 1$ for any $D \neq 0$ and by (iii) we have

$$r(p) \left| \frac{p - (D/p)}{2} \right. .$$

This completes the proof of the lemma.

Let us consider the set of primes p for which $(p, QD) = 1$. For these primes we define a function $g(p)$ by

$$g(p) = \frac{p - (D/p)}{r(p)} .$$

We know from (ii) that $g(p)$ is an integer. In [4] we have shown that the values of $g(p)$ can be arbitrarily large. Now we present an improvement of this result.

Lemma 2: Let $d \geqslant 2$ be an integer and let S be the set of all primes p for which $(p, 2QD) = (Q/p) = 1$, $r(p)|_{[p-(D/p)/2]}$, and

$$p \equiv (D/p) \pmod{d}. \qquad (6)$$

Then the set S is infinite and the function $g(p)$ is unbounded on S. Furthermore, in case $D > 0$, the statements remain valid if the condition $(D/p) = 1$ is also required for the elements of S.

Proof of Lemma 2: Let $m = 8QDdx$ be a natural number, where x is an integer, and suppose that $m > n_0$. Then, as we have seen above, there is a prime p such that $r(p) = m$. Naturally $8QDd|r(p)$ and, by (ii), (6) holds for this p; therefore, Lemma 1 implies that $p \in S$ and, in case $D > 0$, $(D/p) = 1$. This shows that S is an infinite set.

Let us suppose that the function $g(p)$ is bounded on S, i.e., $g(p)$ has only a finite number of distinct values: $t_1, t_2, \ldots, t_s$. Let $m = 8QDdx > n_0$, where x is an integer. We define a number n by

$$n = m(mt_1 + 1)(mt_1 - 1) \cdots (mt_s + 1)(mt_s - 1) + m.$$

As we have seen above, there is a prime p for which $r(p) = n$ and $p \in S$; furthermore, $g(p) = t_i$ for some i ($\leqslant s$) because of our supposition. Hence, by means of the definition of $g(p)$, we have

$$p = r(p) \cdot g(p) + (D/p)$$

$$= m \cdot t_i \cdot \prod_{j=1}^{s} (mt_j + 1)(mt_j - 1) + (mt_i + (D/p)).$$

Thus, p is divisible by $mt_i + (D/p)$, which is impossible since p is a prime. This contradiction proves that the function $g(p)$ is really unbounded on the set S.

3. PROOFS OF THE THEOREMS

Proof of Theorem 1: Let (U_n) be a nondegenerate Lucas sequence and let k (>1) be an integer. We generalize the function $g(p)$ for composite integers n; in the following, $g(n)$ means the ratio

$$g(n) = \frac{n - (D/n)}{r(n)}$$

if n is a natural number and $r(n) \mid (n - (D/n))$.

For an integer s ($\geqslant 1$), we denote by T_s the set of all odd integers n of the form $n = p_1 p_2 \cdots p_s$ for which $r(n) \mid (n - (D/n))/2$, $g(n) > 2$, and $p_1, p_2, \ldots, p_s$ are distinct primes with conditions $(p_i, 2QD) = 1$, $(Q/p_i) = 1$, and $p_i \equiv (D/p_i) \pmod{16QDk}$ for $i = 1, 2, \ldots, s$. Lemma 2, with $d = 16QDk$, shows that T_1 is an infinite set. We shall show by induction on s that T_s is infinite for any $s \geqslant 1$.

Suppose s is a positive integer and $m = p_1 p_2 \cdots p_s \in T_s$ for which $(m - 1)/2 > |QD|n_0$. Then there is a prime q such that $r(q) = (m - (D/m))/2$ and $q \nmid m$ since $g(m) > 2$. For this q, by (iv), we obtain $r(mq) = (m - (D/m))/2$; furthermore, obviously $(q, 2QD) = 1$. Using the definition of T_s, we have

$$m = p_1 p_2 \cdots p_s \equiv (D/p_1) \cdot (D/p_2) \cdots (D/p_s)$$

$$= (D/m) \pmod{16QDk},$$

so $8QD|(m - (D/m))/2 = r(q)$ and, by Lemma 1, $(Q/q) = 1$ follows. But (iii) implies that $(q - (D/q))/2$ is divisible by $r(q)$. Furthermore,

$$\frac{mq - (D/mq)}{2} = \frac{m - (D/m)}{2} \cdot q + \frac{q - (D/q)}{2} \cdot (D/m);$$

therefore,

$$r(mq) \left| \frac{mq - (D/mq)}{2} \right.$$

and

$$g(mq) = \frac{mq - (D/mq)}{r(mq)} = 2 \cdot \frac{mq - (D/mq)}{m - (D/m)} > 2.$$

It can be seen that the congruence

$$q \equiv (D/q) \ (\text{mod } 16QDk)$$

also holds; thus, $n = mq$ is an element of T_{s+1}. It is easy to see that distinct m's determine distinct n's, so T_{s+1} is an infinite set if T_s has infinitely many elements. This proves the statement, since the set T_1 is infinite.

The definition of the set T_s together with (i) imply that each element of T_s is a Euler-Lucas pseudoprime with parameters P and Q; thus, the first part of the theorem is proved.

In case $D > 0$, in consequence of Lemma 1 and Lemma 2, $(D/q) = 1$ if $q \in T_1$ or q is the prime defined above. This implies that every set T_s has infinitely many elements which are products of s distinct primes p of the form

$$p = 16QDky + 1 = kx + 1.$$

This completes the proof of the theorem.

Proof of Theorem 2: As it is known, the terms of nondegenerate sequences (U_n) and (V_n) can be written in the form

$$U_n = \frac{\alpha^n - \beta^n}{\alpha - \beta} \quad \text{and} \quad V_n = \alpha^n + \beta^n,$$

respectively, where α and β are the roots of the polynomial $x^2 - Px + Q$. Using these forms, one can easily see that

$$V_n - Q^{\frac{n-1}{2}} \cdot V_1 = D \cdot U_{\frac{n-1}{2}} \cdot U_{\frac{n+1}{2}} \qquad (7)$$

and

$$U_n - (D/n) \cdot Q^{\frac{n-1}{2}} = U_{\frac{n-(D/n)}{2}} \cdot V_{\frac{n+(D/n)}{2}} \tag{8}$$

for any odd positive integer n.

If $Q = 1$ or $Q = -1$, then $D = P^2 - 4Q > 0$ because α/β is not a root of unity for nondegenerate Lucas sequences. In this case, from Theorem 1 and its proof, it follows that there are infinitely many integers n of the form $n = p_1 p_2 \cdots p_s$ such that $p_1, p_2, \ldots, p_s$ are distinct primes, $n | U_{\frac{n-(D/n)}{2}}$, $p_i \equiv 1 \pmod{4k}$, and

$$(D/p_i) = 1 \quad (i = 1, 2, \ldots, s).$$

For these numbers, we have $n \equiv 1 \pmod 4$ and $(D/n) = 1$. Since $Q = \pm1$, from (7) and (8)

$$U_n \equiv 1 \pmod{n} \quad \text{and} \quad V_n \equiv V_1 \pmod{n}$$

follow. This completes the proof of the theorem.

REFERENCES

[1] Baillie, R., and Wagstaff, S. S., Jr. 'Lucas Pseudo-primes.' *Math. Comp.* **35** (1980):1391–1417.

[2] Duparc, H. J. A. 'Divisibility Properties of Recurring Sequences.' Ph.D. dissertation, Amsterdam, 1963.

[3] P. Erdös. 'On the Converse of Fermat's Theorem.' *Amer. Math. Monthly* **56** (1949):623–624.

[4] Kiss, P., and Phong, B. M. 'On a Function Concerning Second Order Recurrences.' *Ann. Univ. Sci. Budapest Eötvös, sec. Math.* **21** (1978):119–122.

[5] Lehmer, D. H. 'An Extended Theory of Lucas' Function.' *Ann. Math.* **31** (1930):419–448.

[6] Lieuwens, E. 'Fermat Pseudo Primes.' Ph.D. dissertation, Delft, 1971.

[7] Parberry, E. A. 'On Primes and Pseudo-Primes Related to the Fibonacci Sequence.' *The Fibonacci Quarterly* **8**, no. 1 (1970):49–60.

[8] Pomerance, C.; Selfridge, J. L.; and Wagstaff, S. S., Jr. 'The Pseudoprimes to $25 \cdot 10^9$.' *Math. Comp.* **35** (1980):1003–1026.

[9] Rotkiewicz, A. 'On Euler Lehmer Pseudoprimes and Strong Lehmer Pseudoprimes with Parameters L, Q in Arithmetic Progressions.' *Math. Comp.* **39** (1982):239–247.

[10] Rotkiewicz, A.   'On Pseudoprimes with Respect to the
      Lucas Sequences.' *Bull. Acad. Polon. Sci., sér. Sci.
      Math. Astronom. Phys.* **21** (1973):793-797.
[11] Schinzel, A.   'Primitive Divisors of the Expression
      $A^n - B^n$ in Algebraic Number Fields.' *J. reine angev.
      Math.* **268/269** (1974):27-33.
[12] Stewart, C. L.   'Primitive Divisors of Lucas and
      Lehmer Numbers.'   In *Transcendence Theory: Advances
      and Applications*, pp. 79-92.   London-New York-San
      Francisco: Academic Press, 1977.

Joseph Lahr

FIBONACCI AND LUCAS NUMBERS AND THE MORGAN-VOYCE POLYNOMIALS IN LADDER NETWORKS AND IN ELECTRIC LINE THEORY

It is a well-known fact that the same mathematical structures may appear in very different fields of science. The
Fibonacci numbers are an excellent example of this phenomenon and the scientist who in his investigations suddenly is
confronted with these numbers is in the pleasant situation
of being able to find an immense number of relationships and
laws which simplify his work in a significant manner.

Let us now consider the following electric network made
of six half-T sections in which all the elements have the
same resistance R.

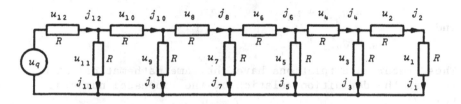

Figure 1. An R ladder consisting of six half-T sections

The determination of the currents in this network is
not complicated if we express the currents as a function of
the so-called reference current j_1. Using Kirchhoff's laws
we obtain relations in which the Fibonacci numbers appear:

$$j_2 = j_1 \qquad j_5 = 5j_1 \qquad j_8 = 21j_1 \qquad j_{11} = 89j_1$$
$$j_3 = 2j_1 \qquad j_6 = 8j_1 \qquad j_9 = 34j_1 \qquad j_{12} = 144j_1$$
$$j_4 = 3j_1 \qquad j_7 = 13j_1 \qquad j_{10} = 55j_1$$

The connection of the reference current j_1 with the source
voltage u_q is given by

$$u_q = u_{11} + u_{12} = (j_{11} + j_{12})R = 233j_1R.$$

141

A. N. Philippou et al. (eds.), Fibonacci Numbers and Their Applications, 141–161.
© 1986 by D. Reidel Publishing Company.

So all the electric magnitudes may be computed by means of the Fibonacci numbers. In general, if the ladder network consists of m half-T sections, i.e., of $2m$ elements, then we obtain, for the reference magnitudes:

$$u_1 = \frac{u_q}{F_{2m+1}} \tag{1}$$

and

$$j_1 = \frac{u_1}{R}. \tag{2}$$

Now we can compute any voltage and any current using the next relations:

$$u_n = F_n u_1; \tag{3}$$

$$j_n = F_n j_1. \tag{4}$$

The appearance of Fibonacci numbers in this context is quite obvious if we consider three consecutive voltages or currents. We have

$$u_n = u_{n-1} + u_{n-2} \tag{5}$$

and

$$j_n = j_{n-1} + j_{n-2}. \tag{6}$$

These recurrence relations have the same mathematical structure as the definition relation of the Fibonacci numbers, namely:

$$F_n = F_{n-1} + F_{n-2} \qquad (F_1 = F_2 = 1). \tag{7}$$

The general term of the Fibonacci numbers,

$$F_n = \frac{1}{\sqrt{5}}\left[\left(\frac{1 + \sqrt{5}}{2}\right)^n - \left(\frac{1 - \sqrt{5}}{2}\right)^n\right], \tag{8}$$

is useful for determining the input impedance Z_i of an infinitely long R ladder network. Proceeding from the input impedance of a ladder consisting of m half-T sections

$$Z_i = \frac{u_q}{j_{2m}} = \frac{F_{2m+1}}{F_{2m}} R, \tag{9}$$

we may establish

$$\lim_{m \to \infty} Z_i = \frac{1 + \sqrt{5}}{2} R = R1,61803\ldots \tag{10}$$

The determination of the electric power losses leads us to the Lucas numbers L_n which satisfy the same recurrence relation but have different starting values:

$$L_n = L_{n-1} + L_{n-2} \qquad (L_1 = 1; \ L_2 = 3). \tag{11}$$

The general term of these numbers also shows the close relationship with the Fibonacci numbers:

$$L_n = \left(\frac{1 + \sqrt{5}}{2}\right)^n + \left(\frac{1 - \sqrt{5}}{2}\right)^n. \tag{12}$$

The total electric power losses P_{tot} may be calculated using the summation formula for the squares of Fibonacci numbers:

$$\sum_{k=1}^{n} F_k^2 = F_n F_{n+1}. \tag{13}$$

Thus,

$$P_{tot} = \sum_{k=1}^{2m} R j_k^2 = j_1^2 R \sum_{k=1}^{2m} F_k^2 = j_1^2 R F_{2m} F_{2m+1}. \tag{14}$$

With the identity

$$F_{2m} F_{2m+1} = \frac{1}{5}(L_{4m+1} - 1) \tag{15}$$

we may write

$$P_{tot} = \frac{j_1^2 R}{5}(L_{4m+1} - 1). \tag{16}$$

Further, it is possible to express the total losses as a function of a sum of two Fibonacci numbers. The relation

$$F_{2n} + F_{2n+2} - (-1)^n = 5 F_n F_{n+1} \tag{17}$$

yields

$$P_{tot} = \frac{j_1^2 R}{5}(F_{4m} + F_{4m+2} - 1). \tag{18}$$

In the practical applications of ladder networks it is often useful to make a difference between the longitudinal and the lateral elements. The determination of power losses in these two groups may be executed with the aid of the following Fibonacci identities:

$$\sum_{k=1}^{n} F_{2k-1}^2 = \frac{1}{5}(F_{4n} + 2n) \tag{19}$$

and

$$\sum_{k=1}^{n} F_{2k}^2 = \frac{1}{5}(F_{4n+2} - 2n - 1). \tag{20}$$

With the notation P_q for the power losses in the lateral elements, we could then write:

$$P_q = \sum_{k=1}^{m} j_{2k-1}^2 R = \frac{j_1^2 R}{5}(F_{4m} + 2m). \tag{21}$$

Similarly, for the losses in the longitudinal elements denoted by P_ℓ:

$$P_\ell = \frac{j_1^2 R}{5}(F_{4m+2} - 2m - 1). \tag{22}$$

The large number of Fibonacci identities could be transformed in relations of currents and voltages if we use the relations (3) and (4) in the form

$$F_n = \frac{j_n}{j_1} \tag{23}$$

and

$$F_n = \frac{u_n}{u_1}. \tag{24}$$

A choice of five identities may illustrate these possibilities:

$$F_{2n+1} = F_{n+1}^2 + F_n^2; \tag{25}$$

$$F_{2n} = F_{n+1}^2 - F_{n-1}^2; \tag{26}$$

$$F_{n+p+1} = F_{n+1}F_{p+1} + F_n F_p; \tag{27}$$

$$F_n F_m - F_{n-k}F_{m+k} = (-1)^{n-k} F_k F_{m+k-n}; \tag{28}$$

$$\pi = 4 \sum_{n=1}^{n=|\infty} \mathrm{Tan}^{-1} \frac{1}{F_{2n+1}}. \tag{29}$$

Transformed in current relations, for instance, one has:

$$j_1 j_{2n+1} = j_{n+1}^2 + j_n^2; \tag{30}$$

$$j_1 j_{2n} = j_{n+1}^2 - j_{n-1}^2; \tag{31}$$

$$j_1 j_{n+p+1} = j_{n+1}j_{p+1} + j_n j_p; \tag{32}$$

$$\mathcal{J}_n \mathcal{J}_m - \mathcal{J}_{n-k} \mathcal{J}_{m+k} = (-1)^{n-k} \mathcal{J}_k \mathcal{J}_{m+k-n} ; \qquad (33)$$

$$\pi = 4 \sum_{n=1}^{n=\infty} \mathrm{Tan}^{-1} \frac{\mathcal{J}_1}{\mathcal{J}_{2n+1}} . \qquad (34)$$

All these considerations justify the statement that currents and voltages in ladder networks are live Fibonacci numbers.

During the study of the Fibonacci numbers and their applications in electrical engineering, it happens that relationships and identities are rediscovered and that they are established in complicated ways. Sometimes a new theoretical concept is found on the basis of invalid arguments. Let me give an example.

The generating function

$$\frac{1}{1-x-x^2} = \sum_{n=1}^{n=\infty} F_n x^{n-1} \qquad (35)$$

yields, with $x = 0.1$ and upon division by 100:

$$\frac{1}{89} = 0.0112358 \qquad (36)$$
$$\qquad\qquad 13$$
$$\qquad\qquad 21$$
$$\qquad\qquad \cdots$$

This interesting result gives the motivation to compute the reciprocals of the Fibonacci numbers:

$$
\begin{array}{llll}
1/F_4 & = 1/3 & = 0.333\ldots & 322 = L_{12} \\
1/F_5 & = 1/5 & = 0.200 & 199 = L_{11} \\
1/F_6 & = 1/8 & = 0.125 & 123 = L_{10} \\
1/F_7 & = 1/13 & = 0.076\ldots & 76 = L_9 \\
1/F_8 & = 1/21 & = 0.047\ldots & 47 = L_8 \qquad (37) \\
1/F_9 & = 1/34 & = 0.029\ldots & 29 = L_7 \\
1/F_{10} & = 1/55 & = 0.018\ldots & 18 = L_6 \\
1/F_{11} & = 1/89 & = 0.011\ldots & 11 = L_5 \\
1/F_{12} & = 1/144 & = 0.0069\ldots & 7 = L_4 \\
\end{array}
$$

We observe that the first three digits after the decimal point represent some resemblance with Lucas numbers. The temptation to apply Lucas identities on the reciprocals

of Fibonacci numbers now exists, and with the well-known
relation $L_n = L_{n-1} + L_{n-2}$ we obtain the conjecture

$$\frac{1}{F_{n-2}} = \frac{1}{F_{n-1}} + \frac{1}{F_n}$$

After some transformations, we obtain the faulty result:

$$F_n^2 = F_n F_{n-1} + F_{n-1}^2.$$

But it is easy to prove that an addition or a subtraction of
1 produces the following correct identity:

$$(-1)^n + F_n^2 = F_n F_{n-1} + F_{n-1}^2. \tag{38}$$

This relation has the property that only two consecutive
Fibonacci numbers appear in it. Referring back to our ladder network we find that this relation also has an important
practical meaning. The transformation of Fibonacci numbers
in voltages, for instance, gives:

$$u_1^2 (-1)^n + u_n^2 = u_n u_{n-1} + u_{n-1}^2. \tag{39}$$

From this equation, it is possible to determine the reference voltage u_1 when two consecutive voltages are known,
regardless of whether n in an even or odd number, since all
voltages are positive:

$$u_1 = \sqrt{(-1)^n (u_n u_{n-1} + u_{n-1}^2 - u_n^2)}. \tag{40}$$

Relation (38) is in fact a special case of a more general
identity concerning the following recurring sequence

$$H_n = aH_{n-1} + bH_{n-2}, \tag{41}$$

namely:

$$(aH_1 H_2 + bH_1^2 - H_2^2)(-b)^{n-2} = aH_n H_{n-1} + bH_{n-1}^2 - H_n^2. \tag{42}$$

This relationship is also of great significance in general
ladder networks.

The principal importance of the study of the electrical
properties of an R ladder network as shown in Figure 1 is to
point out clearly the way to go if one wants to examine more
general ladder networks. That does not mean that there are
no practical applications of the theory just developed. As

an example, I mention a test procedure for computer programs
for the resolution of linear systems of equations.

Suppose we have a ladder network consisting of 50 half-
T sections. The nodal voltages u_1, u_3, u_5, ..., u_{99} may be
the unknown variables in the system of 50 equations. If the
source voltage u_q has a value of F_{101}Volt (F_{101} = 5 73147
84401 38170 84101), then the reference voltage u_1 must have
the value 1 Volt; otherwise, the program contains errors or
the computer arithmetic is not up-to-date.

A further step in the direction of more general ladder
networks consists of making a difference between longitudinal and lateral elements, denoted by R_ℓ and R_q, as shown in
Figure 2.

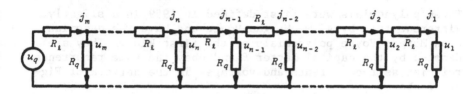

Figure 2. General ladder network made of m half-T sections

It now appears that it is advantageous to treat the
longitudinal and the lateral elements separately. This also
signifies that two groups of subscripts must be defined. If
we restrict our investigation to the currents in the longitudinal elements and the nodal voltages between them, then
the introduction of two different subscripts is superfluous.

The determination of currents and voltages leads to the
realization that the electric properties of this network are
largely dependent on the ratio R_ℓ to R_q so that the introduction of a parameter P is justified, which may be defined
in the following manner:

$$P = 2 + R_\ell/R_q. \tag{43}$$

The application of Kirchhoff's laws now gives the same
recurrence relations for both the currents and the voltages:

$$j_n = Pj_{n-1} - j_{n-2}; \tag{44}$$

$$u_n = Pu_{n-1} - u_{n-2}. \tag{45}$$

If we express the currents as a function of the reference current j_1 and of the parameter P, we obtain the following polynomials:

$$j_2 = Pj_1$$
$$j_3 = (P^2 - 1)j_1$$
$$j_4 = (P^3 - 2P)j_1$$
$$j_5 = (P^4 - 3P^2 + 1)j_1 \qquad (46)$$
$$j_6 = (P^5 - 4P^3 + 3P)j_1$$
$$j_7 = (P^6 - 5P^4 + 6P^2 - 1)j_1$$
$$\ldots$$

These polynomials were first defined in 1959 in a slightly different form by A. M. Morgan-Voyce and have the denomination Morgan-Voyce polynomials of the first kind. They are denoted by the capital letter B and obey the same recurrence relation as the currents and voltages in the network of Figure 2 above:

$$B_n = PB_{n-1} - B_{n-2}. \qquad (47)$$

The initial terms of these polynomials are given by:

$$B_1 = 1 \quad \text{and} \quad B_2 = P. \qquad (48)$$

This definition differs from that chosen by Morgan-Voyce and other authors such as M. N. S. Swamy who start with $B_0 = 1$ and $B_1 = P$. If, from the mathematical point of view, it does not matter if the polynomials start with the subscript 0 or 1, in the practical applications of electrical engineering it is surely more convenient to use the definition proposed here.

The relation between a current j_n and the reference current j_1 then takes the form

$$j_n = B_n j_1. \qquad (49)$$

The determination of the nodal voltages u_n yields the following polynomials, called Morgan-Voyce polynomials of the second kind, denoted by the lower-case letter b:

$$u_2 = (P - 1)u_1$$
$$u_3 = (P^2 - P - 1)u_1$$
$$u_4 = (P^3 - P^2 - 2P + 1)u_1$$
$$u_5 = (P^4 - P^3 - 3P^2 + 2P + 1)u_1 \tag{50}$$
$$u_6 = (P^5 - P^4 - 4P^3 + 3P^2 + 3P - 1)u_1$$
$$u_7 = (P^6 - P^5 - 5P^4 + 4P^3 + 6P^2 - 3P - 1)u_1$$

. . .

The recurrence relation is the same as for the polynomials of the first kind; but the initial terms are different:

$$b_n = Pb_{n-1} - b_{n-2}; \tag{51}$$

$$b_1 = 1 \quad \text{and} \quad b_2 = P - 1. \tag{52}$$

The nodal voltages may now be expressed by

$$u_n = b_n u_1. \tag{53}$$

If the ladder network consists of m half-T sections, then the relationship between the source voltage u_q and the reference voltage u_1 is given by

$$u_1 = \frac{u_q}{b_{m+1}}. \tag{54}$$

For the reference current j_1, one finds immediately

$$j_1 = \frac{u_q}{R_q b_{m+1}}. \tag{55}$$

The input impedance Z_i may now be calculated:

$$Z_i = R_q \frac{b_{m+1}}{B_m}. \tag{56}$$

If the number of sections tends to infinity, then we have

$$\lim_{m \to \infty} Z_i = R_q (X - 1) \tag{57}$$

with

$$X = \frac{P + \sqrt{P^2 - 4}}{2}. \tag{58}$$

The total power losses may be expressed as a product or a sum of Morgan–Voyce polynomials:

$$P_{tot} = j_1^2 R_q b_{m+1} B_m \tag{59}$$

or

$$P_{tot} = j_1^2 \frac{R_\ell}{P^2 - 4}(B_{2m+1} + B_{2m} - 1). \tag{60}$$

For the power losses in the longitudinal elements, we write

$$P_\ell = j_1^2 \frac{R_\ell}{P^2 - 4}(B_{2m+1} - 2m - 1), \tag{61}$$

and for those in the lateral elements we have

$$P_q = j_1^2 \frac{R_\ell}{P^2 - 4}(B_{2m} + 2m). \tag{62}$$

The computation of the most important electric properties of an unloaded ladder network made of m half-T sections may be summarized as in Figure 3.

---

a. DETERMINATION OF THE PARAMETER P

$P = 2 + R_\ell/R_q$

R_ℓ and R_q may be either real or complex.

b. DETERMINATION OF THE MORGAN–VOYCE POLYNOMIALS

First kind: $B_n = PB_{n-1} - B_{n-2}$
with $B_1 = 1$ and $B_2 = P$

Second kind: $b_n = Pb_{n-1} - b_{n-2}$
with $b_1 = 1$ and $b_2 = P - 1$

c. DETERMINATION OF THE ELECTRIC MAGNITUDES

Reference magnitudes: $u_1 = u_q/b_{m+1}$ and $j_1 = u_1/R_q$

Nodal voltages: $u_n = b_n u_1$

Longitudinal currents: $j_n = B_n j_1$

Input impedance: $Z_i = R_q b_{m+1}/B_m$

Figure 3. The electric properties of an unloaded ladder network consisting of m half-T sections

A further step of generalization consists of the introduction of a load Z on the right side of the ladder network,
as shown in Figure 4.

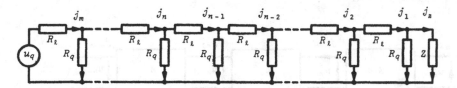

Figure 4. General half-T network with the load Z

The computation of the electric properties of this network may be done according to the flowchart shown in Figure
5 below.

---

a. DETERMINATION OF THE PARAMETER P AND THE MORGAN-VOYCE
POLYNOMIALS ACCORDING TO FIGURE 3

b. DETERMINATION OF THE ELECTRIC MAGNITUDES

Electric magnitudes of the load Z:

$$u_z = \frac{u_q Z}{R_\ell B_m + Z b_{m+1}} \quad \text{and} \quad j_z = \frac{u_z}{Z} \qquad (63)$$

Nodal voltages:

$$u_n = u_z \left(\frac{R_\ell}{Z} B_{n-1} + b_n \right) \qquad (64)$$

Longitudinal currents:

$$j_n = j_z \left(\frac{Z}{R_q} B_n + b_n \right) \qquad (65)$$

Input impedance:

$$Z_i = R_q \frac{R_\ell B_m + Z b_{m+1}}{Z B_m + R_q b_m} \qquad (66)$$

Figure 5. The electric properties of a loaded ladder network
consisting of m half-T sections

Until now we have considered only ladder networks consisting of half-T sections. In order to study all other

possible sections together, it is useful to analyze a ladder
built up of general fourpoles. A fourpole may be characterized by the four a-parameters of the corresponding A-matrix,
as shown in Figure 6.

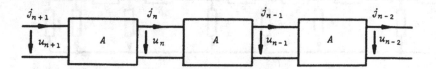

Figure 6. General ladder of fourpoles

The relationships between voltages and currents in any
fourpole of Figure 6 are given by the equations:

$$u_n = a_{11}u_{n-1} + a_{12}j_{n-1}; \tag{67}$$

$$j_n = a_{21}u_{n-1} + a_{22}j_{n-1}. \tag{68}$$

This permits us to establish the following recurrence relations:

$$u_n = (a_{11} + a_{22})u_{n-1} - (a_{11}a_{22} - a_{12}a_{21})u_{n-2} \tag{69}$$

and

$$j_n = (a_{11} + a_{22})j_{n-1} - (a_{11}a_{22} - a_{12}a_{21})j_{n-2}. \tag{70}$$

The coefficient of u_{n-2} and j_{n-2} is the determinant of the
A-matrix, and this determinant always has the value 1 if the
fourpole is constituted of passive elements.

Consequently, the recurrence relations of currents and
voltages have, in ladders of passive fourpoles, the very
simple form:

$$u_n = (a_{11} + a_{22})u_{n-1} - u_{n-2}; \tag{71}$$

$$j_n = (a_{11} + a_{22})j_{n-1} - j_{n-2}. \tag{72}$$

The transition to the Morgan-Voyce polynomials may now be
done without difficulty by setting

$$P = a_{11} + a_{22}. \tag{73}$$

   The determination of the most important electric properties of an unloaded ladder network of fourpoles is summarized in the flowchart of Figure 7. It is remarkable that
only the Morgan-Voyce polynomials of the first kind play a
part in this computation.

---

a. DETERMINATION OF THE PARAMETER P

$$P = a_{11} + a_{22}$$

b. DETERMINATION OF THE MORGAN-VOYCE POLYNOMIALS OF THE
FIRST KIND

$$B_n = PB_{n-1} - B_{n-2} \text{ with } B_1 = 1 \text{ and } B_2 = P$$

c. DETERMINATION OF THE ELECTRIC MAGNITUDES

Reference voltage:

$$u_1 = \frac{u_q}{a_{11}B_m - B_{m-1}} \tag{74}$$

Nodal voltages:

$$u_n = u_1(a_{11}B_{n-1} - B_{n-2}) \tag{75}$$

Longitudinal currents:

$$j_n = u_1 a_{21} B_{n-1} \tag{76}$$

Input impedance:

$$Z_i = \frac{a_{11}B_m - B_{m-1}}{a_{21}B_m} \tag{77}$$

Figure 7. The electric properties of an unloaded ladder network consisting of m passive fourpoles

   If the same ladder as above is loaded with an impedance
Z, then the flowchart of Figure 8 yields all the important
electric properties.

---

a. DETERMINATION OF THE PARAMETER P AND THE MORGAN-VOYCE POLYNOMIALS OF THE FIRST KIND ACCORDING TO FIGURE 7

b. DETERMINATION OF THE ELECTRIC MAGNITUDES

Reference magnitudes:

$$u_1 = \frac{u_q}{\left(a_{11} + \dfrac{a_{12}}{Z}\right)B_m - B_{m-1}} \tag{78}$$

$$j_1 = \frac{u_1}{Z} \tag{79}$$

Nodal voltages:

$$u_n = u_1\left[\left(a_{11} + \frac{a_{12}}{Z}\right)B_{n-1} - B_{n-2}\right] \tag{80}$$

Longitudinal currents:

$$j_n = j_1[(a_{21}Z + a_{22})B_{n-1} - B_{n-2}] \tag{81}$$

Input impedance:

$$Z_i = \frac{(a_{11}Z + a_{12})B_m - ZB_{m-1}}{(a_{21}Z + a_{22})B_m - B_{m-1}} \tag{82}$$

Figure 8. The electric properties of a loaded ladder network consisting of m passive fourpoles

Since the Morgan-Voyce polynomials take such a predominant part in this treatise, it is opportune to give a summary of some mathematical properties.

Definition 1: $B_n = PB_{n-1} - B_{n-2}$

$\qquad\qquad\quad b_n = Pb_{n-1} - b_{n-2}$

$\quad$ *Initial terms:* $B_1 = 1$; $B_2 = P$; $b_1 = 1$; $b_2 = P - 1$

Definition 2: $B_n = (P-1)B_{n-1} + b_{n-1} \tag{83}$

$\qquad\qquad\quad b_n = (P-2)B_{n-1} + b_{n-1} \tag{84}$

$\quad$ *Initial terms:* $B_1 = b_1 = 1$

$$b_n = B_n - B_{n-1} \tag{85}$$

$$-1 + B_n^2 = PB_nB_{n-1} - B_{n-1}^2 \tag{86}$$

$$P - 2 + b_n^2 = Pb_nb_{n-1} - b_{n-1}^2 \tag{87}$$

$$B_{2n-1} = B_n^2 - B_{n-1}^2 \tag{88}$$

$$B_{2n} = B_n(B_{n+1} - B_{n-1}) \tag{89}$$

$$PB_{2n} = B_{n+1}^2 - B_{n-1}^2 \tag{90}$$

$$B_nB_{r-m} = B_rB_{n-m} - B_mB_{n-r} \tag{91}$$

$$B_{n-1}B_{n+1} = B_n^2 - 1 \tag{92}$$

$$b_{n-1}b_{n+1} = b_n^2 + P - 2 \tag{93}$$

$$(P - 2)B_n = (P - 1)b_n - b_{n-1} \tag{94}$$

$$B_{n+1} - B_{n-1} = b_{n+1} + b_n \tag{95}$$

$$b_{m+n} = B_nb_{m+1} - B_{n-1}b_m \tag{96}$$

$$b_{m+n} = B_mb_{n+1} - B_{m-1}b_n \tag{97}$$

$$b_{m+1}B_n - B_mb_{n+1} = b_mB_{n-1} - B_{m-1}b_n \tag{98}$$

$$b_{2n} = B_nb_{n+1} - B_{n-1}b_n \tag{99}$$

$$b_{2n+1} = B_{n+1}b_{n+1} - B_nb_n \tag{100}$$

$$Pb_{2n} = B_{n+1}b_{n+1} - B_{n-1}b_{n-1} \tag{101}$$

$$PB_{2n+1} = B_{n+2}B_{n+1} - B_nB_{n-1} \tag{102}$$

$$b_{2n+1} - b_{2n} = b_{n+1}^2 - b_n^2 \tag{103}$$

$$Pb_{2n+1} = b_{n+2}B_{n+1} - b_nB_{n-1} \tag{104}$$

$$b_nB_{r-m} = b_{n-m}B_r - B_mb_{n-r} \tag{105}$$

$$b_nB_{r-m} = B_{n-m}b_r - b_mB_{n-r} \tag{106}$$

$$b_{n-m}B_r - B_mb_{n-r} = B_{n-m}b_r - b_mB_{n-r} \tag{107}$$

$$b_nB_n - b_{n+1}B_{n-1} = 1 \tag{108}$$

$$B_nb_{n-1} - b_nB_{n-1} = 1 \tag{109}$$

$$B_{m+n-1} = B_mB_n - B_{m-1}B_{n-1} \tag{110}$$

$$b_{m+1}b_m - (P - 2)B_m^2 = 1 \tag{111}$$

$$b_{m+1}^2 - (P - 2)B_mB_{m+1} = 1 \tag{112}$$

$$\sum_{r=1}^{r=n} B_r = \frac{b_{n+1} - 1}{P - 2} \tag{113}$$

$$\sum_{r=1}^{r=n} b_r = B_n \tag{114}$$

$$\sum_{r=1}^{r=n} B_{2r} = B_{n+1}B_n \tag{115}$$

$$\sum_{r=0}^{r=n} B_{2r+1} = B_{n+1}^2 \tag{116}$$

$$\sum_{r=1}^{r=n} b_{2r} = b_{n+1}B_n \tag{117}$$

$$\sum_{r=0}^{r=n} b_{2r+1} = b_{n+1}B_{n+1} \tag{118}$$

$$(P^2 - 4) \sum_{r=1}^{r=n} B_r^2 = B_{2n+1} - 2n - 1 \tag{119}$$

$$(P + 2) \sum_{r=1}^{r=n} b_r^2 = B_{2n} + 2n \tag{120}$$

$$\sum_{r=0}^{r=2n} (-1)^r b_{r+1} = b_{n+1}^2 \tag{121}$$

$$B_n = \sum_{k=0}^{k=n-1} \binom{n+k}{n-k-1}(P-2)^k \tag{122}$$

$$b_n = \sum_{k=0}^{k=n-1} \binom{n-1+k}{n-1-k}(P-2)^k \tag{123}$$

$$B_n = \prod_{k=1}^{k=n-1} \left(P - 2 \cos \frac{k\pi}{n}\right) \quad n \geqslant 2 \tag{124}$$

$$b_n = \prod_{k=1}^{k=n-1} \left[P - 2 \cos \frac{\pi(2k-1)}{2n-1}\right] \quad n \geqslant 2 \tag{125}$$

For the general terms of the Morgan-Voyce polynomials, one must distinguish five different cases if P is a real number.

Case 1. $P > 2$

Auxiliary expressions: X and X'

$$X = \frac{P + \sqrt{P^2 - 4}}{2} \qquad\qquad X' = \frac{P - \sqrt{P^2 - 4}}{2} \tag{126}$$

$$B_n = \frac{X^n - X'^n}{X - X'} \quad \text{or} \quad B_n = \frac{\sinh(n \ln X)}{\sinh(\ln X)} \quad (127)$$

$$b_n = \frac{X^{n-0.5} + X'^{n-0.5}}{X^{0.5} + X'^{0.5}} \quad \text{or} \quad b_n = \frac{\cosh[(n - 0.5)\ln X]}{\cosh(0.5 \ln X)}$$
$$(128)$$

Case 2. $P = 2$

$$B_n = n \qquad\qquad\qquad\qquad\qquad\qquad\qquad (129)$$
$$b_n = 1 \qquad\qquad\qquad\qquad\qquad\qquad\qquad (130)$$

Case 3. $-2 < P < 2$

$$B_n = \frac{\sin[n \cos^{-1}(0.5P)]}{\sin[\cos^{-1}(0.5P)]} \qquad\qquad\qquad (131)$$

$$b_n = \frac{\cos[(n - 0.5)\cos^{-1}(0.5P)]}{\cos[0.5 \cos^{-1}(0.5P)]} \qquad\qquad (132)$$

Case 4. $P = -2$

$$B_n = -n(-1)^n \qquad\qquad\qquad\qquad\qquad (133)$$
$$b_n = (1 - 2n)(-1)^n \qquad\qquad\qquad\qquad (134)$$

Case 5. $P < -2$

Auxiliary expression: P_a

$$P_a = |P| \qquad\qquad\qquad\qquad\qquad\qquad (135)$$

$$B_n = (-1)^{n+1} \frac{\sinh[n \cosh^{-1}(0.5P_a)]}{\sinh[\cosh^{-1}(0.5P_a)]} \qquad (136)$$

$$b_n = (-1)^{n+1} \frac{\sinh[(n - 0.5)\cosh^{-1}(0.5P_a)]}{\sinh[0.5 \cosh^{-1}(0.5P_a)]} \qquad (137)$$

If P is complex, then one must introduce real and imaginary parts of P and of the Morgan-Voyce polynomials:

$$P = a + ib \qquad\qquad\qquad\qquad\qquad\qquad (138)$$

$$B_n = M_n + iN_n \tag{139}$$

$$b_n = V_n + iW_n \tag{140}$$

The recurring determination of the real and imaginary parts is easily calculated using the following relations derived from the second definition of the Morgan-Voyce polynomials [(83), (84)].

$$M_n = (a - 1)M_{n-1} - bN_{n-1} + V_{n-1} \tag{141}$$

$$N_n = (a - 1)N_{n-1} + bM_{n-1} + W_{n-1} \tag{142}$$

$$V_n = (a - 2)M_{n-1} - bN_{n-1} + V_{n-1} \tag{143}$$

$$W_n = (a - 2)N_{n-1} + bM_{n-1} + W_{n-1} \tag{144}$$

Initial terms: $M_1 = V_1 = 1$ and $N_1 = W_1 = 0$ $\quad$ (145)

The following recurrence relation of the fourth order is useful later on for the determination of the general term of the real part of the Morgan-Voyce polynomial of the first kind:

$$M_n = 2aM_{n-1} - (2 + a^2 + b^2)M_{n-2} + 2aM_{n-3} - M_{n-4} \tag{146}$$

The same identity is true for N_n, V_n, and W_n.

The absolute value of B_n, denoted $|B_n|$, may also be expressed by a fourth-order relationship:

$$|B_n|^2 = (a^2 + b^2)|B_{n-1}|^2 + (2 - 2a^2 + 2b^2)|B_{n-2}|^2 +$$
$$+ (a^2 + b^2)|B_{n-3}|^2 - |B_{n-4}|^2. \tag{147}$$

The same identity is true for the absolute value of b_n.

The arguments of the complex morgan-Voyce polynomials of the first kind may be denoted by λ_n. The tangent of this argument

$$\tan \lambda_n = N_n/M_n \tag{148}$$

can be calculated by a nonlinear recurrence relation of the third order,

$$\tan \lambda_n = \frac{(a \tan \lambda_{n-1} + b)h_1 + h_2 \tan \lambda_{n-2}}{(a - b \tan \lambda_{n-1})h_1 + h_2}, \tag{149}$$

with:

$$h_1 = (a - b \tan \lambda_{n-2}) \tan \lambda_{n-3} - a \tan \lambda_{n-2} - b; \tag{150}$$

$$h_2 = \tan \lambda_{n-1} - \tan \lambda_{n-3}. \tag{151}$$

The tangent of the argument ψ_n of the complex Morgan-Voyce polynomials of the second kind can be computed in the same way.

The determination of the general terms of the real and imaginary parts of the complex Morgan-Voyce polynomials is a complicated operation. It seems that the best results are obtained if we introduce the following seven auxiliary expressions: A, B, D, d, E, L, and Z.

$$A = \sqrt[4]{a^4 + b^4 + 2a^2 b^2 - 8a^2 + 8b^2 + 16} \tag{152}$$

$$B = \frac{1}{2} \tan^{-1} \frac{2ab}{a^2 - b^2 - 4} \tag{153}$$

$$D = \frac{1}{2} \sqrt{a^2 + b^2 + A^2 + 2A(a \cos B + b \sin B)} \tag{154}$$

$$d = \ln D \tag{155}$$

$$E = \tan^{-1} \frac{b + A \sin B}{a + A \cos B} \tag{156}$$

$$L = \frac{1}{2} \tan^{-1} \frac{b}{a + 2} \tag{157}$$

$$Z = \sqrt[4]{a^2 + b^2 + 4a + 4} \tag{158}$$

$$B_n = \frac{\sinh[n(d + iE)]}{\sinh(d + iE)} \tag{159}$$

$$b_n = \frac{\cosh[(n - 0.5)(d + iE)]}{\cosh[0.5(d + iE)]} \tag{160}$$

$$M_n = \frac{1}{A}[D^n \cos(nE - B) - D^{-n} \cos(nE + B)] \tag{161}$$

$$N_n = \frac{1}{A}[D^n \sin(nE - B) + D^{-n} \sin(nE + B)] \tag{162}$$

$$\lambda_n = \tan^{-1} \frac{N_n}{M_n} = \tan^{-1}[\coth(nd)\tan(nE)] - B \tag{163}$$

$$|B_n| = \frac{1}{A}\sqrt{2 \; \cosh(2nd) \; - \; 2 \; \cos(2nE)} \qquad (164)$$

$$V_n = \frac{1}{Z}\{D^{n-0.5}\cos[(n - 0.5)E - L] +$$

$$+ \; D^{-(n-0.5)}\cos[(n - 0.5)E + L]\} \qquad (165)$$

$$W_n = \frac{1}{Z}\{D^{n-0.5}\sin[(n - 0.5)E - L] -$$

$$- \; D^{-(n-0.5)}\sin[(n - 0.5)E + L]\} \qquad (166)$$

$$\psi_n = \tan^{-1}\frac{W_n}{V_n} = \tan^{-1}\{\tanh[(n - 0.5)d] \; \times$$

$$\times \; \tan[(n - 0.5)E]\} - L \qquad (167)$$

$$|b_n| = \frac{1}{Z}\sqrt{2 \; \cosh[(2n-1)d] + 2 \; \cos[(2n-1)E]} \qquad (168)$$

With this set of relationships we are in the possession of a strong tool to treat all thinkable problems about any ladder network. A significant profit from the use of Morgan-Voyce polynomials is that in many applications the whole calculations can be done exclusively through the use of the four basic arithmetic operations. Even in the case of the ladder network consisting either solely of inductances and capacitances or of complex impedances, the electric state can be determined by calculating in the domain of real numbers without recourse to usual trigonometric and hyperbolic functions. The problem of representing electric lines by ladder networks is also solved. Thanks to the computer, it does not matter if the line is simulted by 10 or by 100 fourpole sections. The general validity of this method makes it possible to compute electric transmission lines as well as the lines used in modern telecommunication and data processing.

The concepts of characteristic impedance and of propagation constant can be expressed in terms of Morgan-Voyce polynomials. But, ladies and gentlemen, I think I must now stop the formula bombardment, and I thank you very much for your indulgent attention.

REFERENCES

Basin, S. L. 'The Appearance of Fibonacci Numbers and the Q-Matrix in Electrical Network Theory.' *Math. Mag.* **36** (1963):86-97.

Basin, S. L. 'An Application of Continuants.' *Math. Mag.* **37** (1964):83-91.

Lahr, J. H. G. "Theorie elektrischer Leitungen unter Anwendung und Erweiterung der Fibonacci-Funktion.' Dissertation ETH Nr. 6958, Zürich, 1981.

Morgan-Voyce, A. M. 'Ladder Network Analysis Using Fibonacci Numbers.' *I.R.E. Trans. Circuit Theory* **6**, no. 3 (1959): 321-322.

Swamy, M. N. S. "Properties of the Polynomials Defined by Morgan-Voyce.' *The Fibonacci Quarterly* **4**, no. 1 (1966): 73-81.

Swamy, M. N. S. 'More Fibonacci Identities.' *The Fibonacci Quarterly* **4**, no. 4 (1966):369-372.

Swamy, M. N. S. 'Further Properties of Morgan-Voyce Polynomials.' *The Fibonacci Quarterly* **6**, no. 2 (1968):167-175.

Swamy, M. N. S., and Bhattacharyya, B. B. 'A Study of Recurrent Ladders Using the Polynomials Defined by Morgan-Voyce.' *IEEE Trans. on Circuit Theory* **CT-14**, no. 3 (1967).

Br. J. M. Mahon and A. F. Horadam

INFINITE SERIES SUMMATION INVOLVING RECIPROCALS OF PELL POLYNOMIALS

1. INTRODUCTION

Pell polynomials $P_n(x)$ are defined by the recurrence relation

$$P_{n+2}(x) = 2xP_{n+1}(x) + P_n(x), \tag{1.1}$$
$$P_0(x) = 0, \; P_1(x) = 1.$$

Pell-Lucas polynomials $Q_n(x)$ are defined by

$$Q_{n+2}(x) = 2xQ_{n+1}(x) + Q_n(x), \tag{1.2}$$
$$Q_0(x) = 2, \; Q_1(x) = 2x.$$

Binet forms are, respectively,

$$P_n(x) = \frac{\alpha^n - \beta^n}{\alpha - \beta} \tag{1.3}$$

and

$$Q_n(x) = \alpha^n + \beta^n, \tag{1.4}$$

where α and β are the roots of the characteristic equation

$$t^2 - 2xt - 1 = 0 \tag{1.5}$$

so that

$$\begin{cases} \alpha = x + \sqrt{x^2 + 1} \\ \beta = x - \sqrt{x^2 + 1} \end{cases} \text{with } \alpha + \beta = 2x, \; \alpha\beta = -1, \\ \alpha - \beta = 2\sqrt{x^2 + 1}. \tag{1.6}$$

When $x = 1$, the corresponding *Pell numbers* P_n and *Pell-Lucas numbers* Q_n occur. Accordingly, α and β become γ and δ, respectively, the roots of

$$t^2 - 2t - 1 = 0, \tag{1.5'}$$

163

A. N. Philippou et al. (eds.), Fibonacci Numbers and Their Applications, 163–180.
© *1986 by D. Reidel Publishing Company.*

that is,

$$\begin{cases} \gamma = 1 + \sqrt{2} \\ \delta = 1 - \sqrt{2} \end{cases} \quad \text{whence } \gamma + \delta = 2, \; \gamma\delta = -1, \\ \qquad\qquad\qquad\qquad \gamma - \delta = 2\sqrt{2}. \qquad (1.6)'$$

From (1.3), when $x = 1$, we deduce

$$\lim_{n \to \infty} \frac{P_{nk}}{P_{(n+1)k}} = \frac{1}{\gamma^k}. \qquad (1.7)$$

In this paper, we apply to Pell and Pell-Lucas numbers the procedures applied to Fibonacci numbers in [1] and [2], which we also extend and generalize. In addition to the methods employed in [1] and [2], matrix and determinant techniques are used in [5], but not specifically in this paper. Other properties of Pell numbers can be found in [3].

Generally in this paper, $n \geqslant 0$.

Sometimes the adjectival noun "Pell" will be used generically to cover Pell-Lucas properties.

For subsequent specific reference purposes, we append the following identities established in [4] and [5]:

$$P_{m+n}(x) = P_{m-1}(x)P_n(x) + P_m(x)P_{n+1}(x); \qquad (1.8)$$

$$P_{n+r}(x) + P_{n-r}(x) = P_n(x)Q_r(x), \; r \text{ even}; \qquad (1.9)$$

$$P_n^2(x) - P_{n+r}(x)P_{n-r}(x) = (-1)^{n-r}P_r^2(x); \qquad (1.10)$$

$$Q_n^2(x) - Q_{n+r}(x)Q_{n-r}(x) = (-1)^{n+r+1}4(x^2 + 1)P_r^2(x). \qquad (1.11)$$

Additional information, which is necessary to establish the results in the ensuing sections, is detailed in [5] particularly and, generally, also in [4].

Notation for products of polynomials employed in [2] and [5], and used later, is

$$(P_n(x))_6 = \prod_{i=0}^{5} P_{n+i}(x) \qquad (1.12)$$

and

$$(Q_n(x))_6 = \prod_{i=0}^{5} Q_{n+i}(x). \qquad (1.13)$$

Readers will perceive that the notational concepts in (1.12) and (1.13) seem natural extensions of the rising factorial notation

$$(\alpha)_r = \alpha(\alpha + 1)(\alpha + 2) \cdots (\alpha + r - 1). \qquad (1.14)$$

Besides applying the methods in [1] and [2] to Pell polynomials as stated above, we use the notion of finding the sum of a series by means of a reduction formula on a sequence of sums.

Our principal objective in this paper is to obtain sums of infinite series whose terms involve the reciprocals of Pell numbers.

2.   INFINITE SUMMATION OF RECIPROCALS OF PELL POLYNOMIALS

Substituting nr for n in (1.10) and performing the necessary algebraic manipulations, we find that

$$\sum_{r=1}^{n} \frac{(-1)^{(r-1)k}}{P_{kr}(x)P_{(r+1)k}(x)} = \frac{P_{kn}(x)}{P_k^2(x)P_{(n+1)k}(x)} \qquad (2.1)$$

$$\left[= \frac{P_n(x)}{P_{n+1}(x)} \quad \text{when } k = 1 \right].$$

Putting $x = 1$ to obtain the corresponding series for Pell numbers and proceeding to the limiting sum, we calculate by means of (1.3) and (1.7) that

$$\sum_{r=1}^{\infty} \frac{(-1)^{(r-1)k}}{P_{kr}P_{(r+1)k}} = \frac{(\gamma - \delta)^2}{\gamma^k(\gamma^k - \delta^k)^2}. \qquad (2.2)$$

Simplest specializations of this infinite series are

$$\sum_{r=1}^{\infty} \frac{(-1)^{r-1}}{P_r P_{r+1}} = -\delta \qquad (k = 1) \qquad (2.3)$$

and

$$\sum_{r=1}^{\infty} \frac{1}{P_{2r}P_{2(r+1)}} = \frac{\delta^2}{4} \qquad (k = 2). \qquad (2.4)$$

Now

$$\frac{1}{P_n(x)P_{n+1}(x)} - \frac{1}{P_{n+1}(x)P_{n+2}(x)}$$

$$= \frac{P_{n+2}(x) - P_n(x)}{P_n(x)P_{n+1}(x)P_{n+2}(x)}$$

$$= \frac{2x}{P_n(x)P_{n+2}(x)} \qquad \text{by (1.1),}$$

whence, summing,

$$\sum_{r=1}^{n} \frac{1}{P_r(x)P_{r+2}(x)}$$

$$= \frac{1}{2x}\left\{\frac{1}{P_1(x)P_2(x)} - \frac{1}{P_{n+1}(x)P_{n+2}(x)}\right\}. \qquad (2.5)$$

Putting $x = 1$ and letting $n \to \infty$, we obtain

$$\sum_{r=1}^{\infty} \frac{1}{P_r P_{r+2}} = \frac{1}{4}. \qquad (2.6)$$

Observe that the first term of (2.4) is equal to the first two terms of (2.6) minus half the first two terms of (2.3). Likewise, for the second term of (2.4) in relation to the third and fourth terms of (2.3) and of (2.6). Letting $n \to \infty$ in this process, we have the verification that

$$\frac{\delta^2}{4} = \frac{1}{4} + \frac{\delta}{2} \qquad [\text{i.e., } \delta^2 - 2\delta - 1 = 0 \text{ as in } (1.5)'];$$

therefore, (2.6) could have been obtained directly from (2.3) and (2.4).

Next, using (1.11), we derive

$$\sum_{r=1}^{n} \frac{(-1)^{(r-1)k+1}}{Q_{kr}(x)Q_{(r+1)k}(x)} = \frac{-P_{kn}(x)}{P_k(x)Q_k(x)Q_{(n+1)k}(x)} \qquad (2.7)$$

$$\left[= \frac{-P_n(x)}{2xQ_{n+1}(x)} \qquad \text{for } k = 1 \right].$$

whence

$$\sum_{r=1}^{\infty} \frac{(-1)^{(r-1)k}}{Q_{kr}Q_{(r+1)k}} = \frac{1}{\gamma^k(\gamma^{2k} - \delta^{2k})}, \qquad (2.8)$$

from which it ensues that

$$\sum_{r=1}^{\infty} \frac{(-1)^r}{Q_r Q_{r+1}} = \frac{\delta}{\gamma^2 - \delta^2} \qquad (k = 1) \qquad (2.9)$$

and

$$\sum_{r=1}^{\infty} \frac{1}{Q_{2r} Q_{2(r+1)}} = \frac{\delta^2}{\gamma^4 - \delta^4} \qquad (k = 2). \qquad (2.10)$$

A companion result to (2.8) is

$$\sum_{r=1}^{n} \frac{(-1)^{(r-1)k}}{Q_{(r-1)k}(x) Q_{rk}(x)} = \frac{P_{kn}(x)}{2P_k(x) Q_{kn}(x)}. \qquad (2.11)$$

This is more difficult to prove than (2.8) (see [5]). When $x = 1$ and $k = 1$, the infinite sum is, by (1.3), (1.4), and (1.6)′,

$$\sum_{r=1}^{\infty} \frac{(-1)^{r-1}}{Q_{r-1} Q_r} = \frac{1}{2(\gamma - \delta)} \qquad (2.12)$$

$$= \frac{1}{Q_0 Q_1} - \frac{\delta}{\gamma^2 - \delta^2} \qquad \text{from (2.9)}.$$

Finally in this section we may demonstrate, as for (2.5), that

$$\sum_{r=1}^{\infty} \frac{1}{Q_r Q_{r+2}} = \frac{1}{24}. \qquad (2.13)$$

Remarks similar to those made after (2.6) apply here also, on appealing to (2.10), (2.13), and (2.9) in succession with an adjustment of sign in (2.9). [Evidently, $\gamma^2 + \delta^2 = 6$ from (1.6)′, and our *modus operandi* shows that $\delta^2 = 3\delta + \sqrt{2} = 2\delta + 1$ in conformity with (1.5)′.] Consequently, (2.13) could have been produced directly from (2.9) and (2.10).

Various extensions of the above theme present themselves.

Consider, for example, the identity

$$P_{(r+1)k}(x) P_{(r-2)k}(x) - P_{rk}(x) P_{(r-1)k}(x)$$
$$= (-1)^{rk+1} P_k(x) P_{2k}(x), \qquad (2.14)$$

which may be demonstrated by employing the Binet form (1.3) or by determinantal and matrix techniques introduced in [5].

It may be established from (2.14) that:

$$\sum_{r=3}^{n} \frac{(-1)^{rk+1}}{P_{(r-2)k}(x)P_{(r-1)k}(x)}$$

$$= \frac{1}{P_k(x)P_{2k}(x)}\left(\frac{P_{(n+1)k}(x)}{P_{(n-1)k}(x)} - \frac{P_{3k}(x)}{P_k(x)}\right); \qquad (2.15)$$

$$\sum_{r=3}^{n} \frac{(-1)^{rk+1}}{P_{(r-2)k}(x)P_{rk}(x)}$$

$$= \frac{1}{P_k(x)P_{2k}(x)}\left(\frac{P_{(n+1)k}(x)}{P_{nk}(x)} + \frac{P_{nk}(x)}{P_{(n-1)k}(x)} - \right.$$

$$\left. - \frac{P_{3k}(x)}{P_{2k}(x)} - \frac{P_{2k}(x)}{P_k(x)}\right); \qquad (2.16)$$

$$\sum_{r=2}^{n} \frac{(-1)^{rk+1}}{P_{(r-1)k}(x)P_{(r+1)k}(x)}$$

$$= \frac{1}{P_k(x)P_{2k}(x)}\left(\frac{P_k(x)}{P_{2k}(x)} - \frac{P_{nk}(x)}{P_{(n+1)k}(x)} - \frac{P_{(n-1)k}(x)}{P_{nk}(x)}\right). \qquad (2.17)$$

Corresponding limiting sums when $x = 1$ are:

$$\sum_{r=3}^{\infty} \frac{(-1)^{rk+1}}{P_{(r-2)k}P_{(r-1)k}} = \frac{1}{P_k P_{2k}}\left(\gamma^{2k} - \frac{P_{3k}}{P_k}\right); \qquad (2.18)$$

$$\sum_{r=3}^{\infty} \frac{(-1)^{rk+1}}{P_{(r-2)k}P_{rk}} = \frac{1}{P_k P_{2k}}\left(2\gamma^k - \frac{P_{3k}}{P_{2k}} - \frac{P_{2k}}{P_k}\right); \qquad (2.19)$$

$$\sum_{r=3}^{\infty} \frac{(-1)^{rk+1}}{P_{(r-1)k}P_{(r+1)k}} = \frac{1}{P_k P_{2k}}\left(\frac{P_k}{P_{2k}} - 2\gamma^k\right). \qquad (2.20)$$

Other identities like (2.14), the establishment of whose existence and properties parallels those in (2.15)-(2.20), are:

$$Q_{(r+1)k}(x)Q_{(r-2)k}(x) - Q_{rk}(x)Q_{(r-1)k}(x)$$
$$= (-1)^{rk} 4(x^2 + 1)P_k(x)P_{2k}(x); \tag{2.21}$$

$$Q_{(r+1)k}(X)P_{(r-2)k}(x) - Q_{rk}(x)P_{(r-1)k}(x)$$
$$= (-1)^{rk+1}P_k(x)Q_{2k}(x); \tag{2.22}$$

$$P_{(r+1)k}(x)Q_{(r-2)k}(x) - P_{rk}(x)Q_{(r-1)k}(x)$$
$$= (-1)^{rk}\Gamma_k(x)Q_{2k}(x) \tag{2.23}$$

$$P_{(r+1)k}(x)Q_{rk}(x) - P_{rk}(x)Q_{(r+1)k}(x)$$
$$= (-1)^{rk}2P_k(x). \tag{2.24}$$

Moreover, one can produce further similar formulas. Analogously to the proof and properties of (2.14), we derive

$$P_{(r+s+1)k}(x)P_{rk}(x) - P_{(r+s)k}(x)P_{(r+1)k}(x)$$
$$= (-1)^{rk+1}P_k(x)P_{ks}(x), \tag{2.25}$$

and the infinite summation

$$\sum_{r=1}^{\infty} \frac{(-1)^{rk+1}}{P_{rk}P_{(r+1)k}} = \frac{1}{P_k P_{ks}}\left(\gamma^{ks} - \frac{P_{(s+1)k}}{P_k}\right). \tag{2.26}$$

Included in the infinite series which follow from (2.21)–(2.25) are a few whose sums are expressible in alternative, but equivalent, ways.

Identity (2.25) appears to open up still further opportunities for investigation.

Seemingly, the process is endless.

3. MISCELLANEOUS INFINITE SERIES FOR PELL POLYNOMIALS

As there are very many results involving Pell polynomials (see [5]), we think it would be tedious to quote them all. What we do instead is to record the corresponding infinite summations as in [5].

Nonetheless, to get something of the theoretical flavor of the work, we do prove one result for polynomials.

First, we demonstrate that

$$\sum_{r=2}^{n} \frac{P_r(x)}{P_{r-1}(x)P_{r+1}(x)} = \frac{1}{2x}\left\{1 + \frac{1}{2x} - \frac{1}{P_n(x)} - \frac{1}{P_{n+1}(x)}\right\}. \tag{3.1}$$

Proof of (3.1):

$$\frac{P_r(x)}{P_{r-1}(x)P_{r+1}(x)} = \frac{1}{2x}\left(\frac{P_{r+1}(x) - P_{r-1}(x)}{P_{r+1}(x)P_{r-1}(x)}\right) \quad \text{by (1.1)}$$

$$= \frac{1}{2x}\left(\frac{1}{P_{r-1}(x)} - \frac{1}{P_{r+1}(x)}\right).$$

Summation over r from 2 to n yields the required result.

Consequently,

$$\sum_{r=2}^{\infty} \frac{P_r}{P_{r-1}P_{r+1}} = \frac{3}{4}. \tag{3.2}$$

Arguing in a similar way to that in (3.1) gives

$$\sum_{r=1}^{n} \frac{Q_r(x)}{Q_{r-1}(x)Q_{r+1}(x)} = \frac{1}{2x}\left\{\frac{1}{2} + \frac{1}{2x} - \frac{1}{Q_n(x)} - \frac{1}{Q_{n+1}(x)}\right\}, \tag{3.3}$$

whence,

$$\sum_{r=1}^{\infty} \frac{Q_r}{Q_{r-1}Q_{r+1}} = \frac{1}{2}. \tag{3.4}$$

Specializing results proved for Pell-Lucas polynomials in [5], Mahon established the following summation values for selected infinite series involving Pell and Pell-Lucas numbers (there being no significance in the order of listing):

$$\sum_{r=1}^{\infty} \frac{(-1)^r}{P_r P_{r+3}} = \frac{3}{50}(3 + 10\delta); \tag{3.5}$$

$$\sum_{r=1}^{\infty} \frac{(-1)^r}{Q_r Q_{r+3}} = \frac{37 + 63\delta}{840}; \tag{3.6}$$

$$\sum_{r=1}^{\infty} \frac{(-1)^{r-1}Q_{2r+2}}{P_r P_{r+1}Q_{r+1}Q_{r+2}} = \frac{1}{6}; \tag{3.7}$$

$$\sum_{r=1}^{\infty} \frac{(-1)^{r-1}Q_{r+1}}{P_r P_{r+1}P_{r+2}} = \frac{1}{2}; \tag{3.8}$$

$$\sum_{r=1}^{\infty} \frac{(-1)^{r-1} P_{r+1}}{Q_r Q_{r+1} Q_{r+2}} = \frac{1}{96} ; \qquad (3.9)$$

$$\sum_{r=1}^{\infty} \frac{(-1)^{\binom{r+3}{2}} Q_{r+1}}{P_r P_{r+2}} = \frac{3}{2} ; \qquad (3.10)$$

$$\sum_{r=1}^{\infty} \frac{(-1)^{\binom{r+3}{2}} P_{r+1}}{Q_r Q_{r+2}} = \frac{1}{12} ; \qquad (3.11)$$

$$\sum_{r=1}^{\infty} \frac{(-1)^{r-1}}{P_{2r-1} P_{2r+3}} = \frac{1}{30} ; \qquad (3.12)$$

$$\sum_{r=1}^{\infty} \frac{(-1)^{r-1}}{Q_{2r-1} Q_{2r+3}} = \frac{1}{168} ; \qquad (3.13)$$

$$\sum_{r=1}^{\infty} \frac{Q_{r+2}}{P_r P_{r+2} P_{r+4}} = \frac{29}{240} ; \qquad (3.14)$$

$$\sum_{r=1}^{\infty} \frac{P_{r+2}}{Q_r Q_{r+2} Q_{r+4}} = \frac{29}{11424} ; \qquad (3.15)$$

$$\left. \begin{array}{l} \displaystyle\sum_{r=1}^{\infty} \frac{P_{2r+5}}{(P_r)_6} = \frac{1}{1392} \qquad\qquad (3.16) \\[4mm] \displaystyle\sum_{r=1}^{\infty} \frac{P_{2r+5}}{(Q_r)_6} = \frac{1}{535296} \qquad (3.17) \end{array} \right\} \text{[cf. (1.12), (1.13)]}$$

$$\sum_{r=3}^{\infty} \frac{P_{2r}}{P_{r-2}^2 P_{r+2}^2} = \frac{4669}{43200} \qquad (3.18)$$

$$\sum_{r=1}^{\infty} \frac{P_{2r}}{Q_{r-2}^2 Q_{r+2}^2} = \frac{7}{864} \qquad (3.19)$$

$$\sum_{r=1}^{\infty} \frac{(-1)^{r-1} P_{(2k-1)(2r+1)}}{P_{(2k-1)r}^2 P_{(2k-1)(r+1)}^2} = \frac{1}{P_{2k-1}^3} \qquad (3.20)$$
$$= 1 \qquad \text{when } k = 1$$
$$= 1/125 \qquad \text{when } k = 2 ;$$

$$\sum_{r=1}^{\infty} \frac{(-1)^{r-1} P_{(2k-1)(2r+1)}}{Q_{(2k-1)r}^2 Q_{(2k-1)(r+1)}^2} = \frac{1}{8P_{2k-1} Q_{2k-1}^2} \tag{3.21}$$

$$= 1/32 \qquad \text{when } k = 1$$

$$= 1/7840 \quad \text{when } k = 2.$$

The common numerator in (3.14) and (3.15) is a suffi-cient oddity to deserve a comment. In (3.14) it arises from

$$\frac{1}{P_1 P_3} + \frac{1}{P_2 P_4} = \frac{1}{5} + \frac{1}{24},$$

while in (3.15) its origin lies in

$$\frac{1}{Q_1 Q_3} + \frac{1}{Q_2 Q_4} = \frac{1}{28} + \frac{1}{204}.$$

From the nature of the information provided in the last two sections, one can see that the summing of reciprocals offers a large range of possibilities for development. Our brief treatment of some of the opportunities which are available is to be regarded largely as a report on work in progress.

4.   SEQUENCES OF SUMS

Write

$$S(n, k) = \sum_{r=1}^{n} \frac{1}{P_r(x) P_{r+k}(x)} \tag{4.1}$$

We proceed to prove the reduction formula

$$(4x^2 + 2)S(n, 4) - 2S(n, 2)$$

$$= \frac{1}{P_{n+1}(x) P_{n+3}(x)} + \frac{1}{P_{n+2}(x) P_{n+4}(x)} -$$

$$- \frac{1}{P_1(x) P_3(x)} - \frac{1}{P_2(x) P_4(x)}. \tag{4.2}$$

Proof of (4.2): Replace n by $n + 2$ in (1.9) and put $r = 2$. Divide throughout by $P_n(x) P_{n+2}(x) P_{n+4}(x)$. Sum from 1 to n. Accordingly,

$$\sum_{r=1}^{n} \frac{1}{P_r(x)P_{r+2}(x)} - (4x^2 + 2)\sum_{r=1}^{n} \frac{1}{P_r(x)P_{r+4}(x)}$$

$$= -\sum_{r=1}^{n} \frac{1}{P_{r+2}(x)P_{r+4}(x)} .$$

Expressed in terms of (4.1), this is

$$S(n, 2) - (4x^2 + 2)S(n, 4) = -S(n, 2) - \mathscr{P}(x),$$

where $\mathscr{P}(x)$ is the right-hand side of (4.2).
The result follows immediately.

More generally,

$$P_{n+2k}(x) = P_{2k-1}(x)P_n(x) + P_{2k}(x)P_{n+1}(x) \quad \text{by (1.8)}$$
$$= \{P_{2k}(x)P_{n+2}(x) - P_{2k-2}(x)P_n(x)\}/2x$$

by (1.1) used twice.

Algebraic maneuvering as indicated and then summing
give:

$$\sum_{r=1}^{n} \frac{1}{P_r(x)P_{r+2}(x)} - \frac{P_{2k}(x)}{2x}\sum_{r=1}^{n} \frac{1}{P_r(x)P_{r+2k}(x)}$$

$$= -\frac{P_{2k-2}(x)}{2x}\sum_{r=1}^{n} \frac{1}{P_{r+2}(x)P_{r+2k}(x)},$$

whence

$$\frac{P_{2k}(x)}{2x}S(n, 2k) = S(n, 2) + \frac{P_{2k-2}(x)}{2x}S(n, 2k-2) +$$

$$+ \frac{P_{2k-2}(x)}{2x}\left\{\frac{1}{P_{n+1}(x)P_{n+2k-1}(x)} + \frac{1}{P_{n+2}(x)P_{n+2k}(x)} - \right.$$

$$\left. - \frac{1}{P_1(x)P_{2k-1}(x)} - \frac{1}{P_2(x)P_{2k}(x)}\right\}. \tag{4.3}$$

Particular cases which are noted in [5] occur when
$k = 3, 4$. The case $k = 2$ is already recorded in (4.2).

Paralleling the above approach, we may demonstrate (see [5]), beginning with (1.8) with $m = 3$, that

$$P_3(x)S(n, 3) = S(n, 1) - P_2(x)S(n, 2) + P_2(x) \times$$

$$\times \left\{ \frac{1}{P_1(x)P_3(x)} - \frac{1}{P_{n+1}(x)P_{n+3}(x)} \right\}. \quad (4.4)$$

Generally, with $m = 2k + 1$, (1.8) leads to

$$P_{2k+1}(x)S(n, 2k + 1) = S(n, 1) - P_{2k}(x)S(n, 2k) +$$

$$+ P_{2k}(x) \left\{ \frac{1}{P_1(x)P_{2k+1}(x)} - \frac{1}{P_{n+1}(x)P_{n+2k+1}(x)} \right\}. \quad (4.5)$$

Observe, in (4.4) and (4.5), the centrality of the role of $S(n, 1)$.

By (4.3), $S(n, 2k)$ is expressible eventually in terms of $S(n, 2)$, 2hich is known [see (2.5)]. However, by (4.5), $S(n, 2k + 1)$ depends on $S(n, 1)$, which we do not as yet know.

5. USE OF LAMBERT SERIES AND JACOBIAN ELLIPTIC FUNCTIONS

Define the *Lambert series* $L(x)$ by

$$L(x) = \sum_{n=1}^{\infty} \frac{x^n}{1 - x^n}. \quad (5.1)$$

Consider

$$\frac{1}{P_{2n}} = \frac{2\sqrt{2}}{\gamma^{2n} - \delta^{2n}} \qquad \text{from (1.3), (1.6)}' \qquad n \geqslant 1$$

$$= 2\sqrt{2} \frac{\delta^{2n}}{1 - \delta^{4n}} \qquad \because \gamma\delta = -1 \text{ by (1.6)}'$$

$$= 2\sqrt{2} \left(\frac{\delta^{2n}}{1 - \delta^{2n}} - \frac{\delta^{4n}}{1 - \delta^{4n}} \right)$$

$$\therefore \sum_{n=1}^{\infty} \frac{1}{P_{2n}} = 2\sqrt{2} \left(\sum_{n=1}^{\infty} \frac{\delta^{2n}}{1 - \delta^{2n}} - \sum_{n=1}^{\infty} \frac{\delta^{4n}}{1 - \delta^{4n}} \right) \quad \delta^2 < 1$$

$$= 2\sqrt{2}(L(\delta^2) - L(\delta^4)), \qquad \text{by (5.1)}$$

i.e.,

$$\sum_{n=1}^{\infty} \frac{1}{P_{2n}} = 2\sqrt{2}\,(L(3 - 2\sqrt{2}) - L(17 - 12\sqrt{2}))\ \text{by (1.6)}'.$$
(5.2)

Lambert series have thus been found to occur in the infinite summation of the reciprocals of Pell numbers with even subscripts. Next,

$$\frac{1}{Q_{2n-1}} = \frac{1}{\gamma^{2n-1} + \delta^{2n-1}} \qquad \text{from (1.3), (1.6)}'$$

$$= -\frac{\delta^{2n-1}}{1 - \delta^{4n-2}} \qquad \therefore\ \gamma\delta = -1 \text{ by (1.6)}'$$

$$= \frac{\delta^{4n-2}}{1 - \delta^{4n-2}} - \frac{\delta^{2n-1}}{1 - \delta^{2n-1}}$$

$$\therefore \sum_{n=1}^{\infty} \frac{1}{Q_{2n-1}} = \sum_{n=1}^{\infty} \left(\frac{\delta^{4n-2}}{1 - \delta^{4n-2}} - \frac{\delta^{2n-1}}{1 - \delta^{2n-1}} \right)$$

$$= -\sum_{n=1}^{\infty} \frac{\delta^{n}}{1 - \delta^{n}} + 2\sum_{n=1}^{\infty} \frac{\delta^{2n}}{1 - \delta^{2n}} -$$

$$- \sum_{n=1}^{\infty} \frac{\delta^{4n}}{1 - \delta^{4n}} \qquad |\delta| < 1$$

$$= -L(\delta) + 2L(\delta^{2}) - L(\delta^{4}) \qquad \text{by (5.1)}$$

where some algebraic manipulation has been necessary. So, by (1.6)',

$$\sum_{n=1}^{\infty} \frac{1}{Q_{2n-1}} = -L(1 - \sqrt{2}) + 2L(3 - 2\sqrt{2}) - L(17 - 12\sqrt{2}).$$
(5.3)

Thus, Lambert series also occur in the infinite summation of the reciprocals of Pell-Lucas numbers with odd subscripts.

Information concerning Lambert series can be found in [10]. Historically, it is of interest to know that the Lambert series (5.1) played a considerable part in early studies on prime numbers. [If (5.1) is expanded as a power series in x, ;the coefficient of x^n equals the number of divisors of n, i.e., for n prime, the coefficients of x^n are always equal to 2.]

Our attention is now turned to related infinite series whose sum involves Jacobian elliptic functions.

Varying a little the method used in [7] for Fibonacci and Lucas numbers, but preserving for convenience the meaning of the symbol q used therein, let us write, *mutatis mutandis*,

$$q = 3 - 2\sqrt{2} < 1. \qquad (5.4)$$

This is the smaller of the two roots of the equation

$$t^2 - 6t + 1 = 0, \qquad (5.5)$$

which [as may be checked by using (1.1) and (1.2) twice, with $x = 1$] is the characteristic equation of the recurrence relation for both even-subscripted and odd-subscripted Pell and Pell-Lucas sequences. [For Fibonacci and Lucas numbers, the equation corresponding to (5.5) is $t^2 - 3t + 1 = 0.$]

From (1.6)',

$$q = \delta^2, \qquad 1 + q = -2\sqrt{2}\delta. \qquad (5.4)'$$

Now

$$\frac{1}{P_{2n-1}} = \frac{2\sqrt{2}}{\gamma^{2n-1} - \delta^{2n-1}}$$

$$= \frac{(1+q)q^{n-1}}{1 + q^{2n-1}} \qquad \text{by (1.6)', (5.4)'}$$

$$\therefore \sum_{n=1}^{\infty} \frac{1}{P_{2n-1}} = (1+q) \sum_{n=1}^{\infty} \frac{q^{n-1}}{1 + q^{2n-1}}$$

$$= (1+q)\left\{\frac{1}{1+q} + \frac{q}{1+q^3} + \frac{q^2}{1+q^5} + \cdots\right\}$$

$$= (1+q) \cdot \frac{kK}{2\pi\sqrt{q}} \qquad \text{by [9, p. 159]}$$

$$= \frac{1+q}{\sqrt{q}} \cdot \frac{kK}{2\pi}$$

(continued)

$$= \frac{\sqrt{2k}K}{\pi}, \qquad \because \sqrt{q} = -\delta > 0 \qquad (5.6)$$

where the symbols k, K are the standard notation in *Jacobian elliptic functions*.
Furthermore,

$$\frac{1}{Q_{2n}} = \frac{1}{\gamma^{2n} + \delta^{2n}} \qquad \text{by (1.3), (1.6)}'$$

$$= \frac{\delta^{2n}}{1 + \delta^{4n}} \qquad \because \gamma\delta = -1 \text{ by (1.6)}'$$

$$= \frac{q^n}{1 + q^{2n}} \qquad \text{by (5.4)}'$$

$$\therefore \sum_{n=1}^{\infty} \frac{1}{Q_{2n}} = \sum_{n=1}^{\infty} \frac{q^n}{1 + q^{2n}}$$

$$= \frac{q}{1 + q^2} + \frac{q^2}{1 + q^4} + \frac{q^3}{1 + q^6} + \cdots$$

$$= \frac{1}{4}\left(\frac{2K}{\pi} - 1\right) \qquad \text{by [9, p. 159].} \qquad (5.7)$$

Again, as in (5.6), we have found the infinite summation to involve Jacobian elliptic functions.

Eliminating $K/2\pi$ from (5.6) and (5.7), we have (cf. [7, p. 50] for the Fibonacci and Lucas numbers):

$$\frac{1 + \dfrac{1}{5} + \dfrac{1}{29} + \dfrac{1}{169} + \cdots}{\dfrac{1}{4} + \dfrac{1}{6} + \dfrac{1}{34} + \dfrac{1}{198} + \cdots} = 2\sqrt{2}k. \qquad (5.8)$$

Formulas corresponding to (5.2), (5.6), and (5.7) for Fibonacci and Lucas numbers are given in [7], though a formula corresponding to (5.3) has not been noticed in this source. In [11], the formula corresponding to (5.2) for Fibonacci numbers is also given, while that corresponding to (5.6) for Fibonacci numbers is given in terms of *theta-functions*.

Additional information on Jacobian elliptic functions to that on display in [9] can be found in [8] and [13].

6. USE OF GENERALIZED BERNOULLI AND EULER POLYNOMIALS

Following [6], we introduce the *generalized Bernoulli polynomial* $\mathscr{B}_r^{(t)}(x)$ defined by

$$\sum_{r=0}^{\infty} \mathscr{B}_r^{(t)}(x)\frac{n^r}{r!} = e^{nx}\left(\frac{n}{e^n - 1}\right)^t \qquad (6.1)$$

Let

$$C = \frac{\delta}{\gamma} = 2\sqrt{2} - 3 < 0, \qquad \text{by } (1.6)'. \qquad (6.2)$$

Ordinary Bernoulli polynomials $\mathscr{B}_r(x)$ are the special case of (6.1) when $t = 1$.

Information on Bernoulli polynomials, and on Euler polynomials, can be found, for example, in [12].

Reference [6], which deals with a generalization of Fibonacci and Lucas numbers, employs many techniques that parallel those used by Gould, who is concerned with Fibonacci and Lucas numbers. For the Gould and other references, see [6].

Temporarily, write

$$m = n \log C. \qquad (6.3)$$

Since $C < 0$ by (6.2), take the principal value for $\log C$.

Paralleling [6], we have, from $(1.6)'$,

$$\frac{1}{P_n^t} = (-2\sqrt{2})^t \cdot \frac{1}{\gamma^{nt}(C^n - 1)^t} \qquad n \geqslant 1$$

$$= (-2\sqrt{2})^t \cdot \frac{C^{nx}}{(C^x\gamma^t)^n(C^n - 1)^t}$$

$$= (-2\sqrt{2})^t \cdot \frac{1}{m^t(C^x\gamma^t)^n} \cdot e^{mx}\left(\frac{m}{e^m - 1}\right)^t, \qquad \text{by } (6.3),$$

i.e.,

$$\frac{1}{P_n^t} = \frac{(-2\sqrt{2})^t}{m^t(C^x\gamma^t)^n} \sum_{r=0}^{\infty} \mathscr{B}_r^{(t)}(x)\frac{m^r}{r!}, \qquad \text{by } (6.1). \qquad (6.4)$$

Hence, by (6.2) and (6.3), we have the generating function:

$$\sum_{n=1}^{\infty} \frac{1}{P_n^t} y^n = (-2\sqrt{2})^t \sum_{r=0}^{\infty} \mathscr{B}_r^{(t)}(x) \frac{(\log(\delta/\gamma))^{r-t}}{r!} \times$$

$$\times \sum_{n=1}^{\infty} n^{r-t} \left(\frac{y}{\gamma^{t-x}\delta^x} \right)^n.$$
(6.5)

Putting $t = 1$ in (6.4) gives, with (6.2),

$$\frac{1}{P_n} = \frac{-2\sqrt{2}}{(\gamma^{1-x}\delta^x)^n} \sum_{r=0}^{\infty} \mathscr{B}_r(x) \frac{(\log(\delta/\gamma))^{r-1}}{r!} n^{r-1}.$$
(6.6)

In passing, we observe that if we denote

$$\mathscr{P}(x) = \sum_{n=1}^{\infty} \frac{1}{P_n} x^n$$
(6.7)

then

$$\mathscr{P}(\gamma x) - \mathscr{P}(\delta x) = \frac{2\sqrt{2}x}{1-x}$$
(6.8)

on using (1.3) and (1.6)′.

Other results analogous to the Fibonacci-oriented investigations in [6], e.g., recurrence relations for the generalized Bernoulli polynomials and their specialization when $t = 1$, can be duly established.

Finally, designate by $\mathscr{E}_r^{(t)}(x)$ the *generalized Euler polynomial* which we define by

$$\sum_{r=0}^{\infty} \mathscr{E}_r^{(t)}(x) \frac{n^r}{r!} = e^{nx} \left(\frac{2}{e^n+1} \right)^t.$$
(6.9)

Ordinary Euler polynomials $\mathscr{E}_r(x)$ are derived from (6.9) when $t = 1$.

Results for Pell-Lucas numbers Q_n corresponding to those for Pell numbers P_n can then be determined in relation to Euler polynomials. Exploration of this fact is left to the reader.

One of these results, for example, is [cf. (6.6)]:

$$\frac{1}{Q_n} = \frac{1}{2(\gamma^{1-x}\delta^x)^n} \sum_{r=0}^{\infty} \mathscr{E}_r(r) \frac{(\log(\delta/\gamma))^r}{r!} n^r.$$
(6.10)

Obviously, much more could be said about the summation of the reciprocals of Pell-Lucas numbers.

REFERENCES

[1]   Brousseau, A., Br.   'A Summation of Infinite Fibonacci
      Series.'   *The Fibonacci Quarterly* 7, no. 2 (1969)143-
      168.
[2]   Brousseau, A., Br.   'Fibonacci Infinite Series—A
      Research Topic.'   *The Fibonacci Quarterly* 7, no. 3
      (1969):211-217.
[3]   Horadam, A. F.   'Pell Identities.'   *The Fibonacci
      Quarterly* 9, no. 3 (1971):245-263.
[4]   Horadam, A. F., and Mahon, J. M., Br.   'Pell and Pell-
      Lucas Polynomials.'   (In press.)
[5]   Mahon, J. M., Br.   'Pell Polynomials.'   M.A. (Hons.)
      Thesis, The University of New England, Armidale,
      Australia, 1984.
[6]   Shannon, A. G., and Horadam, A. F.   'Reciprocals of
      Generalized Fibonacci Numbers.'   *The Fibonacci Quar-
      terly* 9, no. 3 (1971):299-306, 312.

Historical

[7]   Catalan, E.   'Note sur la Théorie des Fractions Con-
      tinues et sur Certaines Séries.'   *Memoires Academie
      Royale . . . de Belgique* 45 (1883):1-82.
[8]   Cayley, A.   *Elliptic Functions.*   New York: Dover,
      1961 [Bell, 1895].
[9]   Jacobi, C. G. J.   *Fundamenta Nova Theoriae Functionum
      Ellipticarum.*   Volume I of *Gesammelte Werke*, pp. 49-
      239.   Berlin, 1881.
[10] Knopp, K.   *Theory and Application of Infinite Series.*
      New York: Blackie, 1948.
[11] Landau, E.   'Sur la Série des Inverses des Nombres
      de Fibonacci.'   *Bull. Soc. Math.* (France) 27 (1899):
      298-300.
[12] Nörlund, N. E.   *Vorlesungen über Differenzenrechnung.*
      Berlin, 1924.
[13] Whittaker, E. T., and Watson, G. N.   *A Course of
      Modern Analysis.*   C.U.P., 1940.

J. H. McCabe and G. M. Phillips

FIBONACCI AND LUCAS NUMBERS
AND AITKEN ACCELERATION

Given a convergent sequence (x_n), we define an associated sequence (x_n^*) from

$$x_n^* = \frac{x_{n+1}x_{n-1} - x_n^2}{x_{n+1} - 2x_n + x_{n-1}}, \qquad (1)$$

assuming the denominator is nonzero. The derivation of (x_n^*) from (x_n) is known as Aitken acceleration. The process is of practical interest to numerical analysts in those cases in which (x_n^*) converges faster than (x_n), and to the same limit. See, for example, Henrici [4]. For the earliest references, see Aitken [1, 2].

The present paper is not concerned with a practical application of Aitken acceleration, but is rather of number theoretical interest: we present a two-parameter family of sequences (x_n) with the unusual property that each corresponding Aitken sequence (x_n^*) is a *subsequence* of the original sequence (x_n). More precisely, we will find that $x_n^* = x_{2n}$. This family of sequences is defined by

$$x_n = W_{n+1}/W_n, \qquad (2)$$

where

$$W_{n+1} = aW_n - bW_{n-1}, \text{ with } W_1 = 1, W_2 = a, \qquad (3)$$

and where a and b are arbitrary.

The results that follow generalize those of the second author [5], who discussed the special case where $a = -b = 1$; that is, where the sequence (W_n) is the Fibonacci sequence (F_n). (We note in passing that another, even simpler, special case is given by the choice of $a = 2$, $b = 1$, for which $W_n = n$.)

If a or b is zero, the sequence (W_n) is trivial and we exclude such cases from our discussion. Since $x_n = W_{n+1}/W_n$, we also want to ensure that $W_n \neq 0$ for all n. It can be shown that W_n is always nonzero if $b < 0$. If $b > 0$, we need to exclude values of a and b for which

181

A. N. Philippou et al. (eds.), Fibonacci Numbers and Their Applications, 181–184.
© *1986 by D. Reidel Publishing Company.*

$$\cos^{-1}(\tfrac{1}{2}a/\sqrt{b}) = k\pi/n,$$

where k and n are integers satisfying $1 \leqslant k < n$.

The key result is to verify that, with $x_n = W_{n+1}/W_n$ as above,

$$\frac{x_{n+r}x_{n-r} - x_n^2}{x_{n+r} - 2x_n + x_{n-r}} = x_{2n}, \quad 1 \leqslant r < n. \tag{4}$$

We can derive (4) via the intermediate results

$$x_{n+r}x_{n-r} - x_n^2 = \frac{b^{n-r}W_r^2 W_{2n+1}}{W_{n+r}W_{n-r}W_n^2}$$

and

$$x_{n+r} - 2x_n + x_{n-r} = \frac{b^{n-r}W_r^2(W_{n+1} - bW_{n-1})}{W_{n+r}W_{n-r}W_n}$$

which we will not justify here. What is of particular interest is the number of already known identities involving the sequence (W_n) which arise in pursuing the details. These identities, for example

$$W_n^2 - W_{n+m}W_{n-m} = b^{n-m}W_m^2,$$

are even more familiar in the Fibonacci special case.

If we put $r = 1$ in (4), we see from (1) that indeed $x_n^* = x_{2n}$, as we claimed above. However, we can obtain much more from (4). If we replace n by $2n$ in (4) and also put $r = 2$, we see that

$$\frac{x_{n+1}^* x_{n-1}^* - (x_n^*)^2}{x_{n+1}^* - 2x_n^* + x_{n-1}^*} = x_{4n}.$$

In other words, if we apply Aitken acceleration to the (already accelerated) sequence (x_n^*), we again obtain a sub-sequence of the original sequence (x_n). Indeed, we can repeat the Aitken acceleration process as many times as we please, beginning with (x_n) and deriving, in turn, the sequences (x_{2n}), (x_{4n}), (x_{8n}), and so on.

Finally, we mention an obvious extension of the above results. Let us write

$$y_n = Ax_n + B, \tag{5}$$

where $A \neq 0$ and B are any constants, and let (y_n^*) denote the Aitken sequence derived from (y_n). Then it is easily verified from (1) that

$$y_n^* = Ax_n^* + B.$$

It follows that, for any of the sequences (x_n) discussed above,

$$y_n^* = Ax_{2n} + B = y_{2n},$$

so that the subsequence property of Aitken acceleration extends to linear transformations of any of the sequences (x_n) defined above.

In particular, if we choose $x_n = F_{n+1}/F_n$ and put $A = 2$, $B = -1$, we obtain the interesting special case

$$y_n = L_n/F_n, \tag{6}$$

where (L_n) is the Lucas sequence, defined by

$$L_{n+1} = L_n + L_{n-1}, \quad \text{with } L_1 = 1, \; L_2 = 3.$$

If we attempt a direct verification that $y_n^* = y_{2n}$ for the sequence (y_n) defined by (6), we rediscover several well-known identities connecting the Fibonacci and Lucas numbers.

The present paper is concerned with the application of Aitken acceleration to the solution of the quadratic equation $x^2 - ax + b = 0$ by an iterative method. Gill and Murray [3] obtain analogous results for the application of both Newton's method and the secant method to this quadratic equation.

REFERENCES

[1] Aitken, A. C. 'On Bernoulli's Numerical Solution of Algebraic Equations.' *Proc. Roy. Soc. Edin.* **46** (1926): 289.

[2] Aitken, A. C. 'Studies in Practical Mathematics II.' *Proc. Roy. Soc. Edin.* **57** (1937):269.

[3] Gill, J., and Miller, G. 'Newton's Method and Ratios of Fibonacci Numbers.' *The Fibonacci Quarterly* **19**, no. 1 (1981):1.

[4] Henrici, P. *Elements of Numerical Analysis*. New York: John Wiley & Sons, 1964.

[5]   Phillips, G. M.   'Aitken Sequences and Fibonacci Num-
      bers.' *Amer. Math. Monthly* **91** (1984):354.

S. Pethe

ON SEQUENCES HAVING THIRD–ORDER
RECURRENCE RELATIONS

1. INTRODUCTION

Shannon and Horadam [2] studied the third–order recurrence
relation

$$\begin{cases} U_0 = 0, \\ U_n = PU_{n-1} + QU_{n-2} + RU_{n-3} \quad (n \geqslant 4), \end{cases} \qquad (1.1)$$

where they write

$$\{U_n\} = \{J_n\} \text{ when } U_1 = 0, \; U_2 = 1, \; U_3 = P, \qquad (1.2)$$

$$\{U_n\} = \{K_n\} \text{ when } U_1 = 1, \; U_2 = 0, \; U_3 = Q,$$

and

$$\{U_n\} = \{L_n\} \text{ when } U_1 = 0, \; U_2 = 0, \; U_3 = R.$$

Following Barakat [1] and using matrix exponential function,
they then obtained formulas for

$$\sum_{n=0}^{\infty} \frac{J_n}{n!}, \quad \sum_{n=0}^{\infty} \frac{K_n}{n!}, \quad \text{and} \quad \sum_{n=0}^{\infty} \frac{L_n}{n!}$$

in terms of eigenvalues of the matrix

$$X = \begin{bmatrix} a_{11} & a_{12} & a_{13} \\ a_{21} & a_{22} & a_{23} \\ a_{31} & a_{32} & a_{33} \end{bmatrix} \qquad (1.3)$$

with $P = \sum_{i=1}^{3} a_{ii}$, $Q = \sum_{i,j=1}^{3} (a_{ij}a_{ji} - a_{ii}a_{jj})$, $(i \neq j)$, and
$R = \det X$.

A. N. Philippou et al. (eds.), Fibonacci Numbers and Their Applications, 185–192.

The object of this paper is to extend the above results and obtain formulas for

$$\sum_{n=0}^{\infty} \frac{(-1)^n J_{2n+j}}{(2n+j)!}, \quad j = 0, 1,$$

with similar formulas for K_n and L_n. Further, by making use of the Chebychev polynomials of the second kind, we obtain some trigonometric formulas.

2.   SUMMATION FORMULAS FOR J_n, K_n, AND L_n

The characteristic equation of X as defined in (1.3) is

$$\lambda^3 - P\lambda^2 - Q\lambda - R = 0.$$

Therefore, by the Cayley-Hamilton theorem, we have

$$X^3 - PX^2 - QX - RI = 0. \tag{2.1}$$

Lemma 1: We have

$$X^n = J_n X^2 + K_n X + L_n I. \tag{2.2}$$

Proof: It follows from the definitions of J_n, K_n, and L_n that

$$\begin{cases} K_n = QJ_{n-1} + RJ_{n-2}, \text{ and} \\ \\ L_n = RJ_{n-1} \end{cases} \tag{2.3}$$

Formula (2.2) is trivially true for $n = 1$, 2, and 3. For higher values of n it is proved by applying induction and using (2.3).

Following Walton [4], we define, for every square matrix Z, the sine of Z and the cosine of Z by respective power series

$$\sin Z = \sum_{n=0}^{\infty} \frac{(-1)^n Z^{2n+1}}{(2n+1)!}$$

and

$$\cos Z = \sum_{n=0}^{\infty} \frac{(-1)^n Z^{2n}}{(2n)!}.$$

Lemma 2: If $f(t)$ is a polynomial of degree $\leqslant N - 1$, and if $\lambda_1, \lambda_2, \ldots, \lambda_N$ are N distinct eigenvalues of X, then

$$f(X) = \sum_{i=1}^{N} f(\lambda_i) \prod_{\substack{1 \leqslant j \leqslant N \\ j \neq i}} \left[\frac{X - \lambda_j I}{\lambda_i - \lambda_j} \right].$$

Proof: This is Sylvester's matrix interpolation formula [3].

Theorem 1: The following formulas hold for $\{J_n\}$, $\{K_n\}$, and $\{L_n\}$.

$$\sum_{n=0}^{\infty} \frac{(-1)^n J_{2n}}{(2n)!} = \sum_{i=1}^{3} \frac{(\lambda_{i+2} - \lambda_{i+1}) \cos \lambda_i}{D(\lambda)} \tag{2.4}$$

$$\sum_{n=0}^{\infty} \frac{(-1)^n J_{2n+1}}{(2n+1)!} = \sum_{i=1}^{3} \frac{(\lambda_{i+2} - \lambda_{i+1}) \sin \lambda_i}{D(\lambda)} \tag{2.5}$$

$$\sum_{n=0}^{\infty} \frac{(-1)^{n+1} K_{2n}}{(2n)!} = \sum_{i=1}^{3} \frac{(\lambda_{i+2}^2 - \lambda_{i+1}^2) \cos \lambda_i}{D(\lambda)} \tag{2.6}$$

$$\sum_{n=0}^{\infty} \frac{(-1)^{n+1} K_{2n+1}}{(2n+1)!} = \sum_{i=1}^{3} \frac{(\lambda_{i+2}^2 - \lambda_{i+1}^2) \sin \lambda_i}{D(\lambda)} \tag{2.7}$$

$$\sum_{n=0}^{\infty} \frac{(-1)^n L_{2n}}{(2n)!} = \sum_{i=1}^{3} \frac{\lambda_{i+1} \lambda_{i+2} (\lambda_{i+2} - \lambda_{i+1}) \cos \lambda_i}{D(\lambda)} \tag{2.8}$$

$$\sum_{n=0}^{\infty} \frac{(-1)^n L_{2n+1}}{(2n+1)!} = \sum_{i=1}^{3} \frac{\lambda_{i+1} \lambda_{i+2} (\lambda_{i+2} - \lambda_{i+1}) \sin \lambda_i}{D(\lambda)} \tag{2.9}$$

where $D(\lambda) = (\lambda_1 - \lambda_2)(\lambda_2 - \lambda_3)(\lambda_3 - \lambda_1)$ and $\lambda_{i+3} = \lambda_i$, $i = 1, 2$.

Proof: We will prove (2.5), (2.7), and (2.9). The proofs of (2.4), (2.6), and (2.8) are similar.
By Lemma 2, we have

$$\sin X = \sum_{i=1}^{3} \sin(\lambda_i) \prod_{\substack{j=1 \\ j \neq i}}^{3} \left[\frac{X - \lambda_j I}{\lambda_i - \lambda_j} \right], \tag{2.10}$$

where λ_1, λ_2, and λ_3 are the eigenvalues of det. X. Simplifying (2.10), we get

$$\sin X = \frac{1}{D(\lambda)} \sum_{i=1}^{3} \sin(\lambda_i)[(\lambda_{i+2} - \lambda_{i+1})X^2 -$$

$$- (\lambda_{i+2}^2 - \lambda_{i+1}^2)X + \lambda_{i+1}\lambda_{i+2}(\lambda_{i+2} - \lambda_{i+1})I].$$

$$(2.11)$$

Now

$$\sin X = \sum_{n=0}^{\infty} \frac{(-1)^n X^{2n+1}}{(2n+1)!},$$

so, using Lemma 1, we obtain

$$\sin X = \sum_{n=0}^{\infty} \frac{(-1)^n}{(2n+1)!}[J_{2n+1}X^2 + K_{2n+1}X + L_{2n+1}I].$$

$$(2.12)$$

Equating the coefficients of X^2, X, and I in (2.11) and (2.12), we get (2.5), (2.7), and (2.9).

3. SOME SUMMATION FORMULAS INVOLVING TRIGONOMETRIC FUNCTIONS

Consider the sequence given by

$$\begin{cases} V_0(x) = 0, \text{ and} \\ V_n(x) = \dfrac{\sin \frac{1}{2}(n-1)\theta \, \sin \frac{1}{2}n\theta}{\sin(\theta/2) \, \sin \theta}, \quad n \geq 1, \end{cases} \quad (3.1)$$

$$\theta = \arccos x, \ \theta \neq \pm n\pi.$$

Lemma 3: $V_n(x)$ as defined by (3.1) is related to the Chebychev polynomial of the second kind $S_n(x)$ by the following relation.

$$V_n(x) = \sum_{r=1}^{n} S_{r-1}(x), \text{ where} \quad (3.2)$$

$$S_n(x) = \frac{\sin n\theta}{\sin \theta} \quad \text{and} \quad \theta = \arccos x.$$

Proof: The proof is elementary and, hence, omitted here.

Theorem 2: $V_n(x)$ satisfies the following third-order recurrence relation:

$$V_{n+3}(x) = (2x + 1)(V_{n+2}(x) - V_{n+1}(x)) + V_n(x).$$
$$(3.3)$$

Proof: First note that $S_n(x)$ satisfies the second-order recurrence relation

$$S_{n+2}(x) = 2xS_{n+1}(x) - S_n(x).$$
$$(3.4)$$

Now, by (3.2), we have

$$(2x + 1)[V_{n+2}(x) - V_{n+1}(x)] + V_n(x)$$

$$= (2x + 1)\left[\sum_{r=1}^{n+2} S_{r-1}(x) - \sum_{r=1}^{n+1} S_{r-1}(x)\right] + \sum_{r=1}^{n} S_{r-1}(x)$$

$$= (2x + 1)S_{n+1}(x) + \sum_{r=1}^{n} S_{r-1}(x)$$

$$= 2xS_{n+1}(x) - S_n(x) + S_n(x) + S_{n+1}(x) + \sum_{r=1}^{n} S_{r-1}(x).$$

Making use of (3.4) in the above equation we get (3.3). Note that $V_1 = 0$, $V_2 = 1$, and $V_3 = 2x + 1$, so that if we take $P = 2x + 1$, $Q = -(2x + 1)$, and $R = 1$ in (1.1), we see that $\{V_n(x)\}$ is the sequence of type (1.2).

Lemma 4: Let $x = \cos\theta$ and $y = \sin\theta$. The following summation formulas hold.

A. $\sum_{n=0}^{\infty} \frac{\cos n\theta}{n!} = e^x \cos y$

$\sum_{n=0}^{\infty} \frac{(-1)^n \cos n\theta}{n!} = e^{-x} \cos y$

$\sum_{n=0}^{\infty} \frac{\sin n\theta}{n!} = e^x \sin y$

$\sum_{n=0}^{\infty} \frac{(-1)^n \sin n\theta}{n!} = -e^{-x} \sin y$

B. $\displaystyle\sum_{n=0}^{\infty} \frac{\cos 2n\theta}{(2n)!} = \cosh x \cos y$

$\displaystyle\sum_{n=0}^{\infty} \frac{\cos(2n+1)\theta}{(2n+1)!} = \sinh x \cos y$

$\displaystyle\sum_{n=0}^{\infty} \frac{\sin 2n\theta}{(2n)!} = \sinh x \sin y$

$\displaystyle\sum_{n=0}^{\infty} \frac{\sin(2n+1)\theta}{(2n+1)!} = \cosh x \sin y$

C. $\displaystyle\sum_{n=0}^{\infty} \frac{(-1)^n \cos 2n\theta}{(2n)!} = \cos x \cosh y$

$\displaystyle\sum_{n=0}^{\infty} \frac{(-1)^n \cos(2n+1)\theta}{(2n+1)!} = \sin x \cosh y$

$\displaystyle\sum_{n=0}^{\infty} \frac{(-1)^n \sin 2n\theta}{(2n)!} = -\sin x \sinh y$

$\displaystyle\sum_{n=0}^{\infty} \frac{(-1)^n \sin(2n+1)\theta}{(2n+1)!} = \cos x \sinh y$

Proof: Formulas in A are easy to prove. Those in C are proved in [4]. Formulas in B can similarly be proved.

Theorem 3: The following summation formulas for $V_n(x)$ hold.

$$\sum_{n=0}^{\infty} \frac{V_n(x)}{n!} = \frac{1}{4} \operatorname{cosec}^2 \frac{\theta}{2} \sec \frac{\theta}{2}\left[e \cos \frac{\theta}{2} - \right.$$
$$\left. - e^x \cos\left(y - \frac{\theta}{2}\right)\right]; \qquad (3.5)$$

$$\sum_{n=0}^{\infty} \frac{V_{2n}(x)}{(2n)!} = \frac{1}{8} \operatorname{cosec}^2 \frac{\theta}{2} \sec \frac{\theta}{2}\left[2 \cosh 1 \cos \frac{\theta}{2} - \right.$$
$$\left. - e^x \cos\left(y - \frac{\theta}{2}\right) - e^{-x} \cos\left(y + \frac{\theta}{2}\right)\right]; \qquad (3.6)$$

$$\sum_{n=0}^{\infty} \frac{V_{2n+1}(x)}{(2n+1)!} = \frac{1}{8} \operatorname{cosec}^2 \frac{\theta}{2} \sec \frac{\theta}{2}\left[2 \sinh 1 \cos \frac{\theta}{2} - \right.$$
$$\left. - e^x \cos\left(y - \frac{\theta}{2}\right) + e^{-x} \cos\left(y + \frac{\theta}{2}\right)\right]; \qquad (3.7)$$

$$\sum_{n=0}^{\infty} \frac{(-1)^n V_n(x)}{n!} = \frac{1}{4} \operatorname{cosec}^2 \frac{\theta}{2} \sec \frac{\theta}{2} \Big[e^{-1} \cos \frac{\theta}{2} -$$

$$- e^{-x} \cos\Big(y + \frac{\theta}{2}\Big) \Big]; \tag{3.8}$$

$$\sum_{n=0}^{\infty} \frac{(-1)^n V_{2n}(x)}{(2n)!} = \frac{1}{8} \operatorname{cosec}^2 \frac{\theta}{2} \sec \frac{\theta}{2} \Big[2 \cos 1 \cos \frac{\theta}{2} -$$

$$- e^y \cos\Big(x + \frac{\theta}{2}\Big) - e^{-y} \cos\Big(x - \frac{\theta}{2}\Big) \Big];$$
$$\tag{3.9}$$

$$\sum_{n=0}^{\infty} \frac{(-1)^n V_{2n+1}(x)}{(2n+1)!} = \frac{1}{8} \operatorname{cosec}^2 \frac{\theta}{2} \sec \frac{\theta}{2} \Big[2 \sin 1 \cos \frac{\theta}{2} -$$

$$- e^y \sin\Big(x + \frac{\theta}{2}\Big) - e^{-y} \sin\Big(x - \frac{\theta}{2}\Big) \Big].$$
$$\tag{3.10}$$

Proof: We will prove (3.7). The proofs of the others are similar. By trigonometric simplification, it is easy to see that

$$V_n(x) = \frac{1}{4} \operatorname{cosec}^2 \frac{\theta}{2} \Big[1 - \cos n\theta - \tan \frac{\theta}{2} \sin n\theta \Big].$$

Hence

$$\sum_{n=0}^{\infty} \frac{V_{2n+1}(x)}{(2n+1)!} = \frac{1}{4} \operatorname{cosec}^2 \frac{\theta}{2} \Big\{ \sum_{n=0}^{\infty} \frac{1}{(2n+1)!} -$$

$$- \sum_{n=0}^{\infty} \frac{\cos(2n+1)\theta}{(2n+1)!} - \tan \frac{\theta}{2} \sum_{n=0}^{\infty} \frac{\sin(2n+1)\theta}{(2n+1)!} \Big\}.$$
$$\tag{3.11}$$

Using appropriate formulas in B, (3.11) becomes

$$\sum_{n=0}^{\infty} \frac{V_{2n+1}(x)}{(2n+1)!} = \frac{1}{4} \operatorname{cosec}^2 \frac{\theta}{2} \Big\{ \sinh 1 - \sinh x \cos y -$$

$$- \tan \frac{\theta}{2} \cosh x \sin y \Big\}$$

$$= \frac{1}{8} \operatorname{cosec}^2 \frac{\theta}{2} \sec \frac{\theta}{2} \Big\{ 2 \sinh 1 \cos \frac{\theta}{2} -$$

$$- (e^x - e^{-x}) \cos y \cos \frac{\theta}{2} -$$

$$- (e^x + e^{-x}) \sin y \sin \frac{\theta}{2} \Big\}$$

$$= \frac{1}{8} \operatorname{cosec}^2 \frac{\theta}{2} \sec \frac{\theta}{2} \Big\{ 2 \sinh 1 \cos \frac{\theta}{2} -$$

$$- e^x \Big(\cos y \cos \frac{\theta}{2} + \sin y \sin \frac{\theta}{2} \Big) +$$

$$+ e^{-x} \Big(\cos y \cos \frac{\theta}{2} - \sin y \sin \frac{\theta}{2} \Big) \Big\}$$

$$= \frac{1}{8} \operatorname{cosec}^2 \frac{\theta}{2} \sec \frac{\theta}{2} \Big\{ 2 \sinh 1 \cos \frac{\theta}{2} -$$

$$- e^x \cos\Big(y - \frac{\theta}{2} \Big) + e^{-x} \cos\Big(y + \frac{\theta}{2} \Big) \Big\}.$$

Remark: Formulas (3.5)-(3.10) can also be expressed as those involving sine and cosine functions. For example, using (3.1), the left-hand side of (3.7) can be modified as

$$\sum_{n=0}^{\infty} \frac{V_{2n+1}(x)}{(2n+1)!} = \sum_{n=0}^{\infty} \frac{\sin(n\theta)\sin\Big(n + \frac{1}{2}\Big)\theta}{(2n+1)! \ \sin \theta \ \sin \frac{\theta}{2}}.$$

Thus, (3.7) becomes

$$\sum_{n=0}^{\infty} \frac{\sin(n\theta)\sin\Big(n + \frac{1}{2}\Big)\theta}{(2n+1)!}$$

$$= \frac{1}{4}\Big[2 \sinh 1 \cos \frac{\theta}{2} - e^x \cos\Big(y - \frac{\theta}{2} \Big) + e^{-x} \cos\Big(y + \frac{\theta}{2} \Big) \Big],$$

where $x = \cos \theta$ and $y = \sin \theta$.

REFERENCES

[1]   Barakat, R. 'The Matrix Operator e^x and the Lucas Polynomials.' *J. Math. and Phys.* 43 (1964):332-335.

[2]   Shannon, A. G., and Horadam, A. F. 'Some Properties of Third-Order Recurrence Relations.' *The Fibonacci Quarterly* 10, no. 2 (1972):135-145.

[3]   Turnbull, H. W., and Aitken, A. C. *An Introduction to the Theory of Canonical Matrices*, pp. 76-77. New York: Dover, 1961.

[4]   Walton, J. E. 'Lucas Polynomials and Certain Circular Functions of Matrices.' *The Fibonacci Quarterly* 14, no. 1 (1976):83-87.

Attila Pethö

ON THE SOLUTION OF THE EQUATION $G_n = P(x)$

1. INTRODUCTION

Let G_0, G_1, A, $B \in Z$, and $G_{n+1} = AG_n - BG_{n-1}$, for $n \geqslant 1$.
Let α and β denote the roots of the characteristic polynomial $X^2 - AX + B$ of G_n. Finally, let $D = A^2 - 4B$, the discriminant of G_n, $a = G_1 - \beta G_0$, $b = G_1 - \alpha G_0$, and $C = ab$.
The recurrence is called nondegenerated if α/β is not a root
of unity and $C \neq 0$.

Under the assumption of nondegeneracy, T. N. Shorey and
C. L. Stewart [9] proved that all integer solutions x, n,
$q - |x|$, $q \geqslant 2$ of the diophantine equation

$$G_n = dx^q, \quad 0 \neq d \in Z \tag{1}$$

satisfy $\max\{n, |x|, q\} \leqslant C_1$, where C_1 is an effectively com-
putable constant depending only on A, B, G_0, G_1, and d.
Let S denote the set of all nonzero integers composed
of primes p_1, $\ldots$, $p_t \in Z$. Then A. Pethö [6] proved that if
$(A, B) = 1$ then all integer solutions x, q, n, $d - |x|$,
$q \geqslant 2$, $0 \neq d \in S$ of (1) satisfy $\max\{n, |x|, q, d\} \leqslant C_2$,
where C_2 is an effectively computable constant depending
only on A, B, G_0, G_1, p_1, $\ldots$, p_t.
Let $P(x) \in Z[x]$ and denote by $H(P)$ and $\deg(P)$ the
height, i.e., the maximum of the absolute values of the
coefficients of $P(x)$, and the degree of $P(x)$, respectively.
In this paper we are dealing with the more general diophantine equation

$$G_n = dx^q + P(x). \tag{2}$$

This work was written when the author was a visitor at the
Universität zu Köln with the fellowship of the Alexander von
Humboldt-Stiftung.

A. N. Philippou et al. (eds.), Fibonacci Numbers and Their Applications, 193–201.
© 1986 by D. Reidel Publishing Company.

If $|B| = 1$, G_n is nondegenerated, and $P(x)$ is a constant polynomial, then C. L. Stewart [10] was able to prove that (2) has only finitely many effectively computable integer solutions x, n, q with $|x| > 1$, $q > 2$.

This result was extended by I. Nemes and A. Pethö [4]. They proved that if G_n is a nondegenerated recurrence with $|B| = 1$ and $H(P) < h$, $\deg(P) \leqslant \min\{q(1 - \gamma), q - 3\}$, where h and γ denote positive real numbers, then all integer solutions n, x, q with $|x| > 1$ of (2) satisfy $\max\{n, |x|, q\} < C_3$, where C_3 is an effectively computable constant depending only on A, G_0, G_1, d, h, and γ.

For generalizations of this result, we refer to T. N. Shorey and C. L. Stewart [9], P. Kiss [2], and I. Nemes and A Pethö [4].

We shall prove in this paper (Theorem 3) that, if $P(x)$ is a fixed polynomial and $q > \deg(P) + 2$, then (2) has only finitely many effectively computable solutions n, $|x| > 1$, q. This result is best possible in the restriction on q, as was shown in [5].

I. Nemes and A. Pethö [5] have given a necessary condition under which the equation

$$G_n = P(x) = a_k x^k + \cdots + a_0 \tag{3}$$

has infinitely many solutions. They have characterized the solutions in x, too. In Theorem 1 we make more precise this characterization and describe the solutions in n. The result is a generalization of the well-known Skolem-Lech-Mahler theorem.

2.   RESULTS

Let $T(x)$ denote the kth Tshebishef polynomial, i.e., let $T_0(x) = 2$, $T_1(x) = x$, $T_k(x) = x T_{k-1}(x) - T_{k-2}(x)$.

Theorem 1: Let G_n be a linear recurrence with $|B| = 1$, discriminant D, and $P(x) = a_k x^k + \cdots + a_0 \in Z[x]$. Assume that (3) has infinitely many integer solutions n and x.

   (i)   If $\alpha \neq \beta$, then the set of solutions in n is equal to the union of a finite set and a finite number of arithmetical progressions.

   (ii)   If $k > 1$, then the set of integers $k|a_k|x + a_{k-1}$, where x runs through the solutions of (3) is equal to the union of a finite set and a finite number of recurrences with discriminants D_i such that D/D_i are squares of integers.

(iii) If G_n is nondegenerated and $k \geqslant 2$, then

$$P(x) = \epsilon \sqrt{q} T_k \left| \frac{2k|a_k|}{n\sqrt{E}} x + \frac{2a_{k-1}}{n\sqrt{E}} \right|, \qquad (4)$$

where $q = -B^n C/D$, $E = 2(k-1)a_{k-1}^2 - 4ka_k a_{k-2}$ and ϵ, $\eta = \pm 1$.

Remark 1: (iii) and, in a weaker form, (ii) were proved by I. Nemes and A. Pethö in [5].

Remark 2: (i) is a generalization of the well-known Skolem-Lech-Mahler theorem, which is true for more general exponential sums also. It seems to be an interesting question, whether (i) has a generalization to higher-order recurrences or second-order recurrences with $|B| > 1$.

Let R_n be a recursive sequence with $R_0 = 0$, $R_1 = 1$, and with a prime discriminant. Then $C = 1$ and $R_n = (\alpha^n - \beta^n)/\sqrt{D}$. Further, let $R_n^* = \alpha^n + \beta^n$ with the same α and β, then $C^* = -D^* = -D$. The Fibonacci and Lucas sequences satisfy these conditions.

Theorem 2: Put $G_n = R_n$ and assume that (3) has infinitely many integer solutions. Then k is odd, k/n, and there exist integers ℓ_0, ℓ_1 such that $\ell_1 x + \ell_0 = R_{n/k}$.
  Put $G_n = R_n^*$ and assume that (3) has infinitely many integer solutions. Then k/n and there exist integers ℓ_0, ℓ_1 such that either $\ell_1 x + \ell_0 = R_{n/k}$ or $\ell_1 x + \ell_0 = R_{n/k}^*$.

Theorem 3: Let G_n be a nondegenerated recurrence with $|B| = 1$ and $P(x) \in Z[x]$, $0 \neq d \in Z$. There exists an effectively computable constant C_3 depending only on A, G_0, G_1, d, and $P(x)$ such that all integer solutions n, $|x| > 1$, $q > \deg(P) + 2$ of (2) satisfy $\max\{n, |x|, q\} < C_3$.

3. PROOFS

Proof of Theorem 1: Let us assume first that G_n is degenerated but $\alpha \neq \beta$. If $C = 0$, then we may assume $a = G_1 - \beta G_0 = 0$, i.e., $\beta \in Q$, hence $\beta \in Z$, since it is an algebraic integer. By the assumption $|\alpha\beta| = 1$, so $\beta = 1$ or -1, and

$\alpha = -\beta$. Assume now that α/β is a root of unity. Then $|\alpha/\beta| = 1$, and by the assumption $|\alpha\beta| = 1$, so $|\alpha^2| = 1$. Hence, we have seen that if G_n is degenerated, then α and β are roots of unity, consequently G_n is a periodic sequence of integers. This proves (i) and (ii) for degenerated sequences.

From now on we assume that G_n is nondegenerated, and $|\alpha| > |\beta|$. By $|B| = 1$ is $D > 0$, and so are α and β quadratic irrationalities. Hence, G_n tends to infinity. If $\deg(P) = 0$, then (3) has only finitely many solution. Let $\deg(P) = 1$, i.e., $P(x) = a_1 x + a_0$, $a_1 \neq 0$. Then $G_n \pmod{a_1}$ is periodic; hence, the set of solutions n of (3) looks like described in (i).

In the following, we assume that G_n is nondegenerated and $\deg(P) \geqslant 2$. (iii) and, in a weaker form, (ii) were proved by Nemes and Pethö in [5]. To make our argument clear and complete, we give here the sketch of the proof.

Write $G_n = (a\alpha^n - b\beta^n)/(\alpha - \beta)$ and $H_n = a\alpha^n + b\beta^n$. Then $H_n \in Z$ and

$$DG_n^2 + 4CB^n = H_n^2. \tag{5}$$

Let us replace G_n to $P(x)$ in (5), then we have an elliptic equation in the unknowns H_n and x with infinitely many distinct solutions

$$Q(x) = DP(x)^2 + 4CB^n = H_n^2. \tag{6}$$

By the famous theorem of C. L. Siegel [8], (6) has only finitely many solutions, if $Q(x)$ has at least three simple zeros. We have seen that this condition is realized except when $P(x)$ is a solution of the following polynomial equation

$$DP(x)^2 + 4CB = P'(x)^2 R(x), \tag{7}$$

where $R(x) \in Q[x]$ is of degree two without multiple roots. To solve (7), we applied a lemma of Schinzel [7] (Lemma 6, pp. 26-28) and proved (iii).

Finally, we showed that if x and n is a solution of (3) then either $P'(x) = 0$ or there exists an integer z such that

$$D(ka_k x + a_{k-1})^2 - z^2 = DE. \tag{8}$$

From this follows that D/z^2. Let $D = d_1 d_2^2$, where d_1 denotes a quadrat-free integer. d_1 is at least two because of the nondegeneracy. Let $z = d_1 d_2 u$, then (8) is equivalent to the equation

$$(ka_k x + a_{k-1})^2 - d_1 u^2 = E. \tag{9}$$

Let $K = Q(\sqrt{d_1})$ and M be the module of K generated by 1, $\sqrt{d_1}$. Let γ be a fundamental unit in the group of units with norm 1 of the multiplicator ring of M. We may assume without loss of generality that $|\gamma| > 1$. Let δ' denote the conjugate of the element $\delta \in K$. By the theory of norm form equations (see Borevich–Shafarevich [1]), there exist finitely many nonassociated elements $\delta_1, \ldots, \delta_t \in M$ such that the elements of M with norm E are precisely those of form $\delta_i \gamma^h$, where $1 \leqslant i \leqslant t$, and h runs over the integers.

Let x be a solution of (9). Then there exist integers h and i $(1 \leqslant i \leqslant t)$ such that

$$2ka_k x + 2a_{k-1} = \delta_i \gamma^h + \delta_i' \gamma'^h. \tag{10}$$

By (iii),

$$G_n = P(x) = \varepsilon \sqrt{q} T_k ((\delta_i \gamma^h + \delta_i' \gamma'^h)/n\sqrt{E}).$$

But

$$\delta_i \gamma^h \delta_i' \gamma'^h / (n\sqrt{E})^2 = 1,$$

so, by the well-known property of the Tshebishef polynomials

$$G_n = \varepsilon \sqrt{q} \left(\frac{\delta_i}{n\sqrt{E}}\right)^k \gamma^{kh} + \varepsilon \sqrt{q} \left(\frac{\delta_i'}{n\sqrt{E}}\right)^k \gamma'^{kh}. \tag{11}$$

By the assumption, (3) has infinitely many solutions in n, so there exists some i for which (11) has infinitely many solutions. To solve this equation, we apply the following theorem.

Theorem M (M. Mignotte [3]): Suppose that

$$u_m = \sum_{i=1}^{h} P_i(m)\alpha_i^m, \quad v_n = \sum_{i=1}^{h} Q_i(n)\beta_i^n,$$

where the P's and Q's are nonzero polynomials and

$$|\alpha_1| > |\alpha_2| \geqslant \cdots \geqslant |\alpha_m|,$$

$$|\beta_1| > |\beta_2| \geqslant \cdots \geqslant |\beta_n|,$$

$$|\alpha_1| > 1, \quad |\beta_1| > 1.$$

Then

(Mi) There exists an effectively computable integer m_0 such that, for $m > m_0$, the equation

$$u_m = v_n \tag{12}$$

implies $P_1(m)\alpha_1^m = Q_1(n)\beta_1^n$.

(Mii) If (12) has an infinity of solutions, then α_1 and β_1 are multiplicatively dependent.

(Miii) When P_1 and Q_1 are constants, the set of solutions (m, n) of (12) is equal to the union of a finite set and a finite number of arithmetical progressions.

It is clear that (11) fulfils the conditions of (Miii), from which follows (i) at once. Finally, (ii) is a consequence of (i) and (10).

Remark 3: One can deduce from Theorem M that G_n and H_n are closely related to the sequences staying on the right-hand side of (10). Of course, the infinite part of the set of solutions (n, h) of (11) is covered by finitely many arithmetical progressions. Let $m_t = u_1 t + u_2$ and $n_t = v_1 t + v_2$, $t = 1, 2, \ldots$ be a pair of these. With the notation $\hat{\gamma} = \gamma^k$,

$$\tilde{c}_i = \varepsilon\sqrt{q}\,(\delta_i/\eta\sqrt{E})^k \gamma^{v_2 k}, \quad \tilde{d}_i = \varepsilon\sqrt{q}\,(\delta_i'/\varepsilon\sqrt{E})^k \gamma'^{v_2 k},$$

$$\tilde{a} = a\alpha^{u_2}/(\alpha - \beta), \quad \text{and} \quad \tilde{b} = -b\beta^{u_2}/(\alpha - \beta),$$

(11) becomes

$$\tilde{a}\alpha^{u_1 t} + \tilde{b}\beta^{u_1 t} = \tilde{c}\,\hat{\gamma}^{v_1 t} + \tilde{d}\,\hat{\gamma}'^{v_1 t}.$$

Now (Mi) yields $\tilde{a}\alpha^{u_1 t} = \tilde{c}\,\hat{\gamma}^{v_1 t}$. Both $\hat{\gamma}$ and α are units in $Q(\sqrt{d_1})$, so if τ denotes a fundamental unit in this field with $|\tau| > 1$, then there exist integers $U, V > 0$ such that $\hat{\gamma} = \tau^V$ and $\alpha = \tau^U$. Hence

$$\tilde{a}/\tilde{c}_i = \tau^{(v_1 V - u_1 U)t}$$

satisfies for all $t = 0, 1, \ldots$. This means $v_1 V - u_1 U = 0$, $\tilde{a} = \tilde{c}_i$, and $\hat{\gamma}^{v_1} = \alpha^{u_1}$. Finally, $\varepsilon\sqrt{q}/(\eta\sqrt{E})^k \in Q(\sqrt{d_1})$, and its conjugate is either itself or -1 times itself.

Proof of Theorem 2: For $k = 1$, Theorem 2 is trivial. Hence, we may assume $k \geqslant 2$. Both R_n and R_n^{*} are nondegenerated, so $P(x)$ satisfies (4).

Let us first examine the case $G_n = R_n$. Then $q = 1/\varepsilon_n D$ with the notation $-B^n = 1/\varepsilon_n$. In comparing the leading coefficients of (4), we have

$$a_k = \frac{\varepsilon}{\sqrt{\varepsilon_n D}} (2ka_k / n\sqrt{E})^k.$$

This implies that k is odd and $E = \varepsilon_n D F^2$, with an $F \in Z$. So

$$\frac{2ka_k}{n\sqrt{E}} = \sqrt{\varepsilon_n D} \; \frac{2ka_k}{n\varepsilon_n DF} = \sqrt{\varepsilon_n D} \, \ell_1,$$

or, equivalently, $2ka_k = \ell_1 DF$ with an $\ell_1 \in Z$. In comparing the constant terms of (4), we have, analogously, $2a_{k-1} = \ell_0 DF$, with an $\ell_0 \in Z$.

From the proof of Theorem 1, we know that x satisfies (8), which actually has the form

$$D^3 F^2 (\ell_1 x + \ell_0)^2 - z^2 = 4\varepsilon_n D^2 F^2.$$

Hence, z is divisible by DF. Let $z = DFy$, then

$$D(\ell_1 x + \ell_0)^2 - y^2 = 4\varepsilon_n.$$

This means that $\ell_1 x + \ell_0 = R_m$ for an m, and by (iii)

$$R_n = \varepsilon T_k (\sqrt{\varepsilon_n D} \, R_m) / \sqrt{\varepsilon_n D}.$$

From this, k/n follows at once.

We discuss now the case $G_n = R_n^{*}$. If $E = F^2$ or $-F^2$, with an integer F (this satisfies always if k is odd), then we can prove the assertion as in the foregoing case.

Let us assume that $E = fF^2$ with integers f, F, and $|f| \neq 1$ square-free. Then

$$\frac{2ka_k}{n\sqrt{fF}} = \sqrt{f} \; \frac{2ka_k}{nfF} = \sqrt{f} \, \ell_1,$$

and, similarly,

$$\frac{2a_{k-1}}{\sqrt{fF}} = \sqrt{f} \, \ell_0 \quad \text{with integers } \ell_0, \ell_1.$$

After cancellation with F^2 we have, from (8),

$$D^* f^2 (\ell_1 x + \ell_0)^2 - y^2 = 4D^* f. \qquad (13)$$

If f would have a prime divisor p such that $p \nmid D^*$, then p^2 would divide the left-hand side of (13), but it does not divide $4D^* f$. Hence, $f = D^*$ or $-D$ and $\ell_1 x + \ell_0 = R_m$, for some m, finally k/n.

Proof of Theorem 3: Take $\gamma = 1/2$, $\deg(P) = k$, and $H(P) = \max|a_i|$. By the theorem of Nemes and Pethö [4], there exists an effectively computable constant C_2 depending only on A, G_0, G_1, k, and $H(P)$ such that all integral solutions n, $|x| > 1$, $q > \max\{k + 3, 2k\}$ of (2) satisfy

$$\max\{n, |x|, q\} < C_2.$$

If $k \leqslant 3$, then we have nothing to prove. Hence, we may assume $k > 3$, or, equivalently, $2k > k + 3$. We shall see that if $k + 3 < q < 2k$, then (2) has finitely many solutions.

Let us assume that there exists a q_0 with $k + 3 < q_0 < 2k$ such that (2) has infinitely many solutions. Then, by (iii),a the polynomial

$$Q(x) = dx^{q_0} + P(x) = dx^{q_0} + a_k x^k + \cdots + a_0$$

fulfils (4). Actually $E = 0$, but $2q_0 d \neq 0$; therefore, $Q(x)$ cannot have the form (4).

REFERENCES

[1] Borevich, Z. I., and Shafarevich, I. R. *Number Theory*. 2nd ed. New York and London: Academic Press, 1967.

[2] Kiss, P. 'Differences of the Terms of Linear Recurrences.' (To appear.)

[3] Mignotte, M. 'Intersection des images de certaines suites recurrentes lineaires.' *Theor. Comput. Sci.* 7 (1978):117-121.

[4] Nemes, I., and Pethö, A. 'Polynomial Values in Linear Recurrences I.' *Publ. Math. Debrecen* 31 (1984):229-233.

[5] Nemes, I., and Pethö, A. 'Polynomial Values in Linear Recurrences II.' *F. Number Theory* (to appear).

[6]   Pethö, A.   'Perfect Powers in Second Order Linear
      Recurrences.'   *J. Number Theory* **15** (1982):5-13.
[7]   Schinzel, A.   *Selected Topics on Polynomials*.   Ann
      Arbor: University of Michigan Press, 1982.
[8]   Siegel, C. L.   'The Integer Solutions of the Equation
      $y^2 = ax^n + bx^{n-1} + \cdots + k$.'   *J. London Math. Soc.* **1**
      (1926):66-68.
[9]   Shorey, T. N., and Stewart, C. L.   'On the Diophantine
      Equation $ax^{2t} + bx\, y + cy^{2t} = d$ and Pure Powers in
      Recurrences.'   *Math. Scand.* **52** (1983):24-36.
[10] Stewart, C. L.   'On Some Diophantine Equations and
      Related Linear Recurrence Sequences.'   *Seminaire
      Delange-Pisot-Poitou Theorie des Nomb.* (1980-1981):317-
      321.

Andreas N. Philippou

DISTRIBUTIONS AND FIBONACCI POLYNOMIALS OF ORDER k, LONGEST RUNS, AND RELIABILITY OF CONSECUTIVE-k-OUT-OF-n : F SYSTEMS

1. INTRODUCTION AND SUMMARY

Unless otherwise stated, in this paper k is a fixed positive integer, n_i $(1 \leqslant i \leqslant k)$ and n are nonnegative integers as specified, p and x are real numbers in the intervals $(0, 1)$ and $(0, \infty)$, respectively, $q = 1 - p$, and $[x]$ denotes the greatest integer in x. Motivated by the work of Philippou and Muwafi [19], Philippou, Georghiou, and Philippou [21] introduced the geometric distribution of order k, and derived from its study the negative binomial of order k and the Poisson of order k. The latter two were briefly studied by Philippou [14, 15], who also obtained the compound Poisson of order k. Aki, Kuboki, and Hirano [1] further studied the first three distributions mentioned above, introduced the logarithmic distribution of order k, and discussed the meaning of the order of the distributions. Various graphs of them are given by Hirano, Kuboki, Aki, and Kuribayashi [10]. In a forthcoming paper, Philippou and Makri [18] obtained the binomial distribution of order k and stated several open problems. We review briefly all these distributions in Section 2. In Section 3 we review some results of Philippou, Georghiou, and Philippou [22, 23] on Fibonacci and Fibonacci-type polynomials of order k, in conjunction with the longest success runs [17]. Finally, in Section 4 we review some results on the reliability of the consecutive-k-out-of-n : F system [4, 6, 7, 26] and derive three new formulas for its reliability, two of which are closed. We also give a new simple proof of two excellent recursive formulas of Shanthikumar [26], obtain a relation between the Fibonacci polynomials of order k and the entries of the Pascal triangle of order k, and present a partial solution to a problem of [18].

This research has been supported through a grant of the Greek Ministry of Research and Technology.

A. N. Philippou et al. (eds.), Fibonacci Numbers and Their Applications, 203–227.
© *1986 by D. Reidel Publishing Company.*

2. DISTRIBUTIONS OF ORDER k

Let N_k denote the number of Bernoulli trials until the occurrence of the kth consecutive success ($k \geqslant 2$). Shane [25] found a formula for $P(N_k = n)$ ($n \geqslant k$), as well as for $P(N_k \leqslant x)$ ($x \geqslant k$), in terms of the polynacci polynomials of order k in p, and Turner [28] derived another formula for $P(N_k = n)$, in terms of the Pascal triangles of order k. Philippou and Muwafi [19] reconsidered the problem and derived the following closed and simpler formula, in terms of the multinomial coefficients:

$$P(N_k = n + k)$$

$$= p^{n+k} \sum_{\substack{n_1, \ldots, n_k \ni \\ n_1 + 2n_2 + \cdots + kn_k = n}} \binom{n_1 + \cdots + n_k}{n_1, \ldots, n_k} \left(\frac{q}{p}\right)^{n_1 + \cdots + n_k},$$

for $n \geqslant 0$.   (2.1)

The proof of (2.1) is immediate by combinatorial arguments, upon noting that a typical element of the event ($N_k = n + k$) is an arrangement

$$x_1 x_2 \cdots x_{n_1 + \cdots + n_k} \underbrace{ss \cdots s}_{k},$$

such that n_1 of the x's are $e_1 = f$, n_2 of the x's are $e_2 = sf$, $\ldots$, n_k of the x's are $e_k = \underbrace{ss \cdots s}_{k-1} f$, and

$$n_1 + 2n_2 + \cdots + kn_k = n.$$

It should be noted that (2.1) is valid for $k = 1$ also.
    We now consider the following.

Definition 2.1: The sequence $\{f_n^{(k)}\}_{n=0}^{\infty}$ is said to be the Fibonacci sequence of order k if

$$f_0^{(k)} = 0, \quad f_1^{(k)} = 1,$$

and

$$f_n^{(k)} = \begin{cases} f_{n-1}^{(k)} + \cdots + f_1^{(k)} & \text{if } 2 \leqslant n \leqslant k + 1 \\ f_{n-1}^{(k)} + \cdots + f_{n-k}^{(k)} & \text{if } n \geqslant k + 2. \end{cases}$$

Then

$$f_{n+1}^{(k)} = \sum_{\substack{n_1, \ldots, n_k \ni \\ n_1 + 2n_2 + \cdots + kn_k = n}} \binom{n_1 + \cdots + n_k}{n_1, \ldots, n_k}, \text{ for } n \geq 0,$$

$$(2.2)$$

and

$$f_{n+1}^{(k)} = 2^n \sum_{i=0}^{[n/(k+1)]} (-1)^i \binom{n - ki}{i} 2^{-(k+1)i} -$$

$$- 2^{n-1} \sum_{i=0}^{[(n-1)/(k+1)]} (-1)^i \binom{n - 1 - ki}{i} 2^{-(k+1)i},$$

for $n \geq 1$.   (2.3)

Formula (2.2) was obtained in [19] by combinatorial arguments and induction. Both formulas (2.2) and (2.3) were derived by Philippou [13] by means of the generating function of $\{f_n^{(k)}\}_{n=0}^{\infty}$ and the multinomial and binomial theorem, respectively.

Let $\{F_n\}_{n=0}^{\infty}$ and $\{T_n\}_{n=0}^{\infty}$ denote the Fibonacci and Tribonacci sequence, respectively. Then, for $k = 2$ and $k = 3$, respectively, (2.2) reduces to

$$F_{n+1} = \sum_{i=0}^{[n/2]} \binom{n - i}{i}, \quad n \geq 0,$$ (2.4)

and

$$T_{n+1} = \sum_{i=0}^{[n/2]} \sum_{j=0}^{[(n-2i)/3]} \binom{i + j}{i} \binom{n - i - 2j}{i + j}, \quad n \geq 0.$$

$$(2.5)$$

The first formula is well known. The second was first noticed in [19].

Setting $p = 1/2$ in (2.1), we obtain by means of (2.2) the following relationship between probability and the Fibonacci sequence of order k:

$$P(N_k = n + k | p = 1/2) = f_{n+1}^{(k)}/2^{n+k}, \quad n \geq 0.$$ (2.6)

Motivated by the work of Philippou and Muwafi [19], Philippou, Georghiou, and Philippou [21] defined and studied the geometric distribution of order k, initiating the introduction and study of other distributions of order k.

Definition 2.2 [21]: A random variable X is said to have the geometric distribution of order k with parameter p, to be denoted by $G_k(p)$, if

$$P(X = n)$$

$$= p^n \sum_{\substack{n_1, \ldots, n_k \ni \\ n_1 + 2n_2 + \cdots + kn_k = n - k}} \binom{n_1 + \cdots + n_k}{n_1, \ldots, n_k} \left(\frac{q}{p}\right)^{n_1 + \cdots + n_k},$$

for $n \geqslant k$.

Definition 2.3 [15, 21]: A random variable X is said to have the negative binomial distribution of order k with parameter vector (r, p) $(r \geqslant 1$, integer), to be denoted by $NB_k(r, p)$, if

$$P(X = n)$$

$$= p^n \sum_{\substack{n_1, \ldots, n_k \ni \\ n_1 + 2n_2 + \cdots + kn_k = n - kr}} \binom{n_1 + \cdots + n_k + r - 1}{n_1, \ldots, n_k, r - 1} \times$$

$$\times \left(\frac{q}{p}\right)^{n_1 + \cdots + n_k}, \text{ for } n \geqslant kr.$$

Definition 2.4 [14, 21]: A random variable X is said to have the Poisson distribution of order k with parameter λ $(\lambda > 0)$, to be denoted by $P_k(\lambda)$, if

$$P(X = n) = e^{-k\lambda} \sum_{\substack{n_1, \ldots, n_k \ni \\ n_1 + 2n_2 + \cdots + kn_k = n}} \frac{\lambda^{n_1 + \cdots + n_k}}{n_1! \cdots n_k!},$$

for $n \geqslant 0$.

Definition 2.5 [14]: A random variable X is said to have the compound Poisson distribution of order k with parameter vector (r, α) $(r \geqslant 1$, integer and $0 < \alpha < \infty)$, to be denoted by $CP_k(r, \alpha)$, if

$$P(X = n)$$

$$= \left(\frac{\alpha}{k + \alpha}\right)^r \sum_{\substack{n_1, \ldots, n_k \ni \\ n_1 + 2n_2 + \cdots + kn_k = n}} \binom{n_1 + \cdots + n_k + r - 1}{n_1, \ldots, n_k, r - 1} \times$$

$$\times \left(\frac{1}{k + \alpha}\right)^{n_1 + \cdots + n_k}, \text{ for } n \geqslant 0.$$

Definition 2.6 [1]: A random variable X is said to have the logarithmic series distribution with parameter p, to be denoted by $LS_k(p)$, if

$$P(X = n)$$

$$= \frac{p^n}{-k \log p} \sum_{\substack{n_1, \ldots, n_k \ni \\ n_1 + 2n_2 + \cdots + kn_k = n}} \frac{(n_1 + \cdots + n_k - 1)!}{n_1! \cdots n_k!} \times$$

$$\times \left(\frac{q}{p}\right)^{n_1 + \cdots + n_k}, \text{ for } n \geq 1.$$

Definition 2.7 [18]: A random variable X is said to have the binomial distribution of order k with parameter vector (n, p), to be denoted by $B_k(n, p)$, if

$$P(X = x)$$

$$= p^n \sum_{i=0}^{k-1} \sum_{x_1, \ldots, x_k} \binom{x_1 + \cdots + x_k + x}{x_1, \ldots, x_k, x} \left(\frac{q}{p}\right)^{x_1 + \cdots + x_k},$$

for $x = 0, 1, \ldots, [n/k]$,

where the summation is taken over all nonnegative integers $x_1, \ldots, x_k$ such that $x_1 + 2x_2 + \cdots + kx_k = n - i - kx$.

If X is given by any one of the Definitions 2.2–2.6, it can be shown by means of the transformation $n_i = m_i$ $(1 \leq i \leq k)$ and $n = m + \sum_{i=1}^{k} (i - 1)m_i$ that $\sum_n P(X = n) = 1$, which shows that its probability generating function $g_X(s)$ exists for $|s| \leq 1$. The following probability generating functions can be obtained by the same transformation. The means and variances follow immediately by straightforward differentiation of $g_X(s)$.

Proposition 2.1 [21]: Let X be a random variable distributed as $G_k(p)$. Then,

(a) $g_X(s) = p^k s^k (1 - ps)/(1 - s + qp^k s^{k+1})$, $|s| \leq 1$;

(b) $E(X) = (1 - p^k)/qp^k$ and

$$\sigma^2(X) = \{1 - (2k + 1)qp^k \doteq p^{2k+1}\}/q^2 p^{2k}.$$

Remark 2.1: Feller [8] considered the random variable N_k (not X) in the context of renewal theory and derived $g_{N_k}(s)$

as well as $E(N_k)$ and $\sigma^2(N_k)$ by developing two recurrences on the probabilities $p_n = P(N_k = n)$ and $u_n = P$ (a success run of length k occurs at the nth trial). The probability generating function $g_{N_k}(s)$ of N_k was also obtained by Shane [25], who employed for this his own formula for p_n.

Proposition 2.2 [15]: Let X be a random variable distributed as $NB_k(r, p)$. Then,

   (a) $g_X(s) = \{p^k s^k (1 - ps)/(1 - s + qp^k s^{k+1})\}^r$, $|s| \leqslant 1$;

   (b) $E(X) = r(1 - p^k)/qp^k$ and

$$\sigma^2(X) = r\{1 - (2k + 1)qp^k - p^{2k+1}\}/q^2 p^{2k}.$$

Proposition 2.3 [14]: Let X be a random variable distributed as $P_k(\lambda)$. Then

   (a) $g_X(s) = \exp\left\{-\lambda\left(k - \sum_{i=1}^{k} s^i\right)\right\}$, $|s| \leqslant 1$;

   (b) $E(X) = \dfrac{k(k + 1)}{2}\lambda$ and $\sigma^2(X) = \dfrac{k(k + 1)(2k + 1)}{6}\lambda$.

Proposition 2.4 [1]: Let X be a random variable distributed as $LS_k(p)$. Then,

   (a) $g_X(s) = (-k \log p)^{-1}\log\{(1 - ps)/1 - s + pq^k s^{k+1}\}$, $|s| \leqslant 1$;

   (b) $E(X) = (1 - p^k - kqp^k)/qp^k(-k \log p)$;

   (c) $E(X^2) = \{1 - p^{2k+1} - (2k + 1)qp^k\}/q^2 p^{2k}(-k \log p)$.

Proposition 2.5 [14]: Let X be a random variable distributed as $CP_k(r, \alpha)$. Then

   (a) $g_X(s) = \left\{1 + \alpha^{-1}\left(k - \sum_{i=1}^{k} s^i\right)\right\}^{-r}$, $|s| \leqslant 1$.

   (b) $E(X) = k(k + 1)r/2\alpha$ and

$$\sigma^2(X) = \{k(k + 1)(2k + 1)r/6\alpha\} + k^2(k + 1)^2 r/4\alpha^2.$$

Problem 2.1 [18]: Let X be a random variable distributed as $B_k(n, p)$.

(a)   Show that $\displaystyle\sum_{x=0}^{[n/k]} P(X = x) = 1$.

(b)   Evaluate $g_X(s)$, $E(X)$, and $\sigma^2(X)$.

Assuming (a), we shall give a partial solution of (b) in Section 4.

How do the distributions of order k arise? The following theorems provide some answers.

Theorem 2.1 [21, 19]: Let N_k be a random variable denoting the number of Bernoulli trials until the occurrence of the kth consecutive success. Then N_k is distributed as $G_k(p)$.

Theorem 2.2 [21]: Let $X_1, \ldots, X_r$ be independent random variables distributed as $G_k(p)$ and set $Y_r = X_1 + \cdots + X_r$. Then Y_r is distributed as $NB_k(r, p)$.

Theorem 2.3 [15]: Let $T_{k,r}$ be a random variable denoting the number of Bernoulli trials until the occurrence of the rth kth consecutive success. Then $T_{k,r}$ is distributed as $NB_k(r, p)$.

Theorem 2.4 [21]: Let X_r $(r \geqslant 1)$ and X be random variables distributed as $NB_k(r, p)$ and $P_k(\lambda)$, respectively. Then

$$P(X_r - kr = n) \underset{r \to \infty}{\to} P(X = n), \quad n \geqslant 0.$$

Theorem 2.5 [14]: Let X and Λ be two random variables such that $X|\Lambda = \lambda$ is distributed as $P_k(\lambda)$ and

$$f_\Lambda(\lambda) = \frac{\alpha^r}{\Gamma(r)} \lambda^{r-1} e^{-\alpha\lambda}, \quad \lambda > 0.$$

Then X is distributed as $CP_k(r, \alpha)$.

Theorem 2.6 [18]: Let $N_n^{(k)}$ be a random variable denoting the number of success runs of length k in n $(\geqslant 1)$ Bernoulli trials. Then

$$P(N_n^{(k)} = x)$$

$$= p^n \sum_{i=0}^{k-1} \sum_{x_1, \ldots, x_k} \binom{x_1 + \cdots + x_k + x}{x_1, \ldots, x_k, x} \left(\frac{q}{p}\right)^{x_1 + \cdots + x_k},$$

$$x = 0, 1, \ldots, [n/k],$$

where the summation is over all nonnegative integers $x_1, \ldots, x_k$ such that $x_1 + 2x_2 + \cdots + kx_k = n - i - kx$.

Remark 2.2: Theorem 2.6 has been obtained independently by Hirano [9].

The next two theorems follow immediately by means of Propositions 2.3(a) and 2.5(a).

Theorem 2.7 [14]: Let X_i ($1 \leqslant i \leqslant m$) be independent random variables distributed as $P_k(\lambda_i)$, and set $X = X_1 + \cdots + X_m$ and $\lambda = \lambda_1 + \cdots + \lambda_m$. Then X is distributed as $P_k(\lambda)$.

Theorem 2.8 [14]: Let X_i ($1 \leqslant i \leqslant m$) be independent random variables distributed as $CP_k(r_i, \alpha)$, and set $X = X_1 + \cdots + X_m$ and $r = r_1 + \cdots + r_m$. Then X is distributed as $CP_k(r, \alpha)$.

It is well known, due to Raikov [24], that if X and Y are independent random variables and $X + Y$ has a Poisson distribution, then each one of them has a Poisson distribution. Does this result of Raikov carry over to $P_k(\lambda)$? The answer is trivially positive for $k = 1$. It is negative for $k \geqslant 2$ because of the following theorem.

Theorem 2.9 [16]: Let X be a random variable distributed as $P_{k-1}(\lambda)$, $k \geqslant 2$, and let Y be a random variable distributed independently of X with probability function

$$P(Y = y) = e^{-\lambda} \lambda^{y/k} / (y/k)!, \; y = 0, k, 2k, \ldots .$$

Then $X + Y$ is distributed as $P_k(\lambda)$.

Theorem 2.9 follows directly from Lemma 2.1 below, which holds true by means of Definition 2.4.

Lemma 2.1: Let X_i ($1 \leqslant i \leqslant k$) be independent random variables distributed as $P(\lambda)$. Then X is distributed as $P_k(\lambda)$ if and only if

$$X = \sum_{i=1}^{k} i X_i.$$

As another consequence of Lemma 2.1, one may obtain Proposition 2.3 directly, by means of simple expectation properties.

The probability generating function $g_X(s)$ of a random variable X distributed as $P_k(\lambda)$ can be written as

$$g_X(s) = \exp\left\{-\lambda k \left(1 - \frac{1}{k}\sum_{i=1}^{k} s^i\right)\right\}$$

$$= g_N(g_{X_1}(s)) = g_{S_N}(s), \qquad (2.7)$$

where X_i $(i \geqslant 1)$ are independent random variables, distributed uniformly on the integers $1, \ldots, k$ and independently of the random variable N which is distributed as $P(k\lambda)$, and $S_N = X_1 + \cdots + X_N$. We have thus established another derivation of $P_k(\lambda)$, the following.

Theorem 2.10 [30]: Let X_i $(i \geqslant 1)$ be independent random variables, distributed uniformly on the integers $1, \ldots, k$ and independently of the random variable N which is distributed as $P(k\lambda)$, and set $S_N = X_1 + \cdots + X_N$. Then S_N is distributed as $P_k(\lambda)$.

Now let X_i $(i \geqslant 1)$ and S_N be as above, but let N be distributed as $B(n, p)$. Then

$$g_{S_N}(s) = g_N(g_{X_1}(s)) = \left(q + p\frac{1}{k}\sum_{i=1}^{k} s^i\right)^n. \qquad (2.8)$$

It follows from (2.8) that

$$\sum_{m=0}^{kn} s^m P(S_N = m)$$

$$= \left(q + \frac{p}{k}\sum_{i=1}^{k} s^i\right)^n$$

$$= \sum_{\substack{n_0, \ldots, n_k \geqslant \\ n_0 + \cdots + n_k = n}} \binom{n}{n_0, \ldots, n_k} q^{n_0}\left(\frac{p}{k}\right)^{n_1 + \cdots + n_k} s^{n_1 + 2n_2 + \cdots + kn_k}$$

$$= q^n \sum_{\substack{n_1, \ldots, n_k \geqslant \\ n_1 + \cdots + n_k = n}} \binom{n}{n_1, \ldots, n_k, \ n - n_1 - \cdots - n_k} \times$$

$$\times \left(\frac{p}{kq}\right)^{n_1 + \cdots + n_k} s^{n_1 + 2n_2 + \cdots + kn_k}$$

(continued)

$$= \sum_{m=0}^{kn} s^m q^n \sum_{\substack{n_1, \ldots, n_k \ni \\ n_1 + 2n_2 + \cdots + kn_k = m}} \binom{n}{n_1, \ldots, n_k, \, n - n_1 - \cdots - n_k} \times$$

$$\times \left(\frac{p}{kq}\right)^{n_1 + \cdots + n_k}.$$

Therefore,

$$P(S_N = m)$$

$$= q^n \sum_{\substack{n_1, \ldots, n_k \ni \\ n_1 + 2n_2 + \cdots + kn_k = m}} \binom{n}{n_1, \ldots, n_k, \, n - n_1 - \cdots - n_k} \times$$

$$\times \left(\frac{p}{kq}\right)^{n_1 + \cdots + n_k}, \quad m = 0, 1, \ldots, kn. \tag{2.9}$$

The probability distribution defined by (2.9) is another
binomial distribution of order k, which reduces to $B(n, p)$
for $k = 1$. It is obviously a generalized $B(n, p)$ distribution, generalized by the discrete uniform on the integers
$1, \ldots, k$. Denote it by $\tilde{B}_k(n, p)$, and let X_i $(i = 1, 2)$ be
independently distributed as $\tilde{B}_k(n_i, p)$. Then (2.8) implies
that $X_1 + X_2$ is distributed as $\tilde{B}_k(n_1 + n_2, p)$. It also
implies that $\tilde{B}_k(n, p) \to P_k(\lambda k^{-1})$, as $p \to 0$ and $n \to \infty$ with
$np \to \lambda$. The mean and variance of S_N may be obtained
directly from the definition of S_N or from (2.8). They are

$$E(S_N) = \frac{(k + 1)np}{2}$$

and $\tag{2.10}$

$$\sigma^2(S_N) = \frac{(k + 1)np\{3(k + 1)q + k - 1\}}{12}.$$

3. FIBONACCI POLYNOMIALS OF ORDER k, FIBONACCI-TYPE POLYNOMIALS OF ORDER k, AND LONGEST RUNS

Consider the following definitions.

Definition 3.1 [22]: The sequence of polynomials $\{f_n^{(k)}(x)\}_{n=0}^{\infty}$
is said to be the sequence of Fibonacci polynomials of order
k if $f_0^{(k)}(x) = 0$, $f_1^{(k)}(x) = 1$, and

$$f_n^{(k)}(x) = \begin{cases} \displaystyle\sum_{i=1}^{n} x^{k-i} f_{n-i}^{(k)}(x) & \text{if } 2 \leqslant n \leqslant k+1 \\[2em] \displaystyle\sum_{i=1}^{k} x^{k-i} f_{n-i}^{(k)}(x) & \text{if } n \geqslant k+2. \end{cases}$$

If $f_n^{(r)}(x) = 0$ for $-(r-2) \leqslant n \leqslant -1$, Hoggatt and Bicknell [11] call $R_n(x) = f_n^{(r)}(x)$ $(n \geqslant -(r-2))$ r-bonacci polynomials.

Denoting by $F_n(x)$, $f_n^{(k)}$, and $P_n^{(k)}$, respectively, the Fibonacci polynomials [27], the Fibonacci numbers of order k, and the Pell numbers of order k [20], it follows from Definition 3.1 that

$$f_n^{(2)}(x) = F_n(x); \quad f_n^{(k)}(1) = f_n^{(k)}; \quad f_n^{(k)}(2) = P_n^{(k)}. \quad (3.1)$$

Definition 3.2 [23]: The sequence of polynomials $\{F_n^{(k)}(x)\}_{n=0}^{\infty}$ is said to be the sequence of Fibonacci-type polynomials of order k if $F_0^{(k)}(x) = 0$, $F_1^{(k)}(x) = 1$, and

$$F_n^{(k)}(x) = \begin{cases} x\{F_{n-1}^{(k)}(x) + \cdots + F_1^{(k)}(x)\} & \text{if } 2 \leqslant n \leqslant k+1 \\[1em] x\{F_{n-1}^{(k)}(x) + \cdots + F_{n-k}^{(k)}(x)\} & \text{if } n \geqslant k+2. \end{cases}$$

It follows from Definition 3.2 that

$$F_n^{(2)}(1) = F_n \quad \text{and} \quad F_n^{(k)}(1) = f_n^{(k)}, \quad (3.2)$$

which indicates that the first definition is more rich than the second. It may be noted, however, that Definition 3.2 is quite appropriate when dealing with longest runs [17].

The following two theorems give expansions of $f_n^{(k)}(x)$ and $F_n^{(k)}(x)$ in terms of the multinomial and binomial coefficients. Compared with the expansions of Hoggatt and Bicknell [11], which were derived in terms of the elements of the left-justified k-nomial triangle, ours may be considered better.

Theorem 3.1 [22]: Let $\{f_n^{(k)}(x)\}_{n=0}^{\infty}$ be the Fibonacci polynomials of order k. Then

(a) $f_{n+1}^{(k)}(x)$

$$= \sum_{\substack{n_1, \ldots, n_k \ni \\ n_1 + 2n_2 + \cdots + kn_k = n}} \binom{n_1 + \cdots + n_k}{n_1, \ldots, n_k} x^{k(n_1 + \cdots + n_k) - n},$$

for $n \geqslant 0$;

(b) $f_{n+1}^{(k)}(x)$

$$= \left(\frac{1 + x^k}{x}\right)^n \sum_{i=0}^{[n/(k+1)]} (-1)^i \binom{n - ki}{i} x^{ki} (1 + x^k)^{-(k+1)i} -$$

$$- \frac{1}{x}\left(\frac{1 + x^k}{x}\right)^{n-1} \sum_{i=0}^{[(n-1)/(k+1)]} (-1)^i \times$$

$$\times \binom{n - 1 - ki}{i} x^{ki} (1 + x^k)^{-(k+1)i}, \text{ for } n \geqslant 1.$$

Theorem 3.2 [23]: Let $\{F_n^{(k)}(x)\}_{n=0}^{\infty}$ be the Fibonacci-type polynomials of order k. Then

(a) $F_{n+1}^{(k)}(x) = \sum_{\substack{n_1, \ldots, n_k \ni \\ n_1 + 2n_2 + \cdots + kn_k = n}} \binom{n_1 + \cdots + n_k}{n_1, \ldots, n_k} x^{n_1 + \cdots + n_k},$

for $n \geqslant 0$;

(b) $F_{n+1}^{(k)}(x)$

$$= (1 + x)^n \sum_{i=0}^{[n/(k+1)]} (-1)^i \binom{n - ki}{i} x^i (1 + x)^{-(k+1)i} -$$

$$- (1 + x)^{n-1} \sum_{i=0}^{[(n-1)/(k+1)]} (-1)^i \times$$

$$\times \binom{n - 1 - ki}{i} x^i (1 + x)^{-(k+1)i}, \text{ for } n \geqslant 1.$$

The formulas (2.2) and (2.3) follow as a corollary of either Theorem 3.1 or 3.2. We also have the following corollary, by means of Definition 2.2 and Theorems 3.1(a) and 3.2(a).

Corollary 3.1 [22, 23]: Let X be a random variable distributed as $G_k(p)$, and let $\{f_n^{(k)}(x)\}_{n=0}^{\infty}$ and $\{F_n^{(k)}(x)\}_{n=0}^{\infty}$, respectively, be the Fibonacci polynomials of order k and the Fibonacci-type polynomials of order k. Then

(a) $P(X = n + k) = p^{n+k}(q/p)^{n/k} f_{n+1}^{(k)}((q/p)^{1/k})$, $n \geqslant 0$;

(b) $P(X = n + k) = p^{n+k} F_{n+1}^{(k)}(q/p)$, $n \geqslant 0$.

We now give a simple proof of the main result of Uppuluri and Patil [29], which was implicit in the work of Philippou, Georghiou, and Philippou [22].

Corollary 3.2: Let X be as above. Then

$$P(X = n + k)$$

$$= p^k \sum_{i=0}^{[n/(k+1)]} (-1)^i \binom{n - ki}{i} (qp^k)^i -$$

$$- p^{k+1} \sum_{i=0}^{[(n-1)/(k+1)]} (-1)^i \binom{n - 1 - ki}{i} (qp^k)^i, \quad n \geqslant 1.$$

Proof: The proof is an immediate consequence of either Corollary 3.1(a) and Theorem 3.1(b) [with $x = (q/p)^{1/k}$], or Corollary 3.1(b) and Theorem 3.2(b) (with $x = q/p$).

Corollary 3.3: Let $\{f_n^{(k)}(x)\}_{n=0}^{\infty}$ and $\{F_n^{(k)}(x)\}_{n=0}^{\infty}$ be the sequences, respectively, of Fibonacci polynomials of order k and Fibonacci-type polynomials of order k. Then

$$F_{n+1}^{(k)}(x) = x^{n/k} f_{n+1}^{(k)}(x^{1/k}), \quad n \geqslant 0.$$

Proof: The proof is an immediate consequence of Corollary 3.1.

Corollary 3.4 [22]: Let $\{F_n(x)\}_{n=0}^{\infty}$ and $\{P_n^{(k)}\}_{n=0}^{\infty}$ be the sequences, respectively, of Fibonacci polynomials and Pell numbers of order k. Then

(a) $F_{n+1}(x) = \sum_{i=0}^{[n/2]} \binom{n - i}{i} x^{n-2i}$, for $n \geqslant 0$;

(b) $P_{n+1}^{(k)} = \sum_{\substack{n_1, \ldots, n_k \ni \\ n_1 + 2n_2 + \cdots + kn_k = n}} \binom{n_1 + \cdots + n_k}{n_1, \ldots, n_k} 2^{k(n_1 + \cdots + n_k) - n}$,

for $n \geqslant 0$;

(c) $P_{n+1}^{(k)}$

$$= \left(\frac{1 + 2^k}{2}\right)^n \sum_{i=0}^{[n/(k+1)]} (-1)^i \binom{n - ki}{i} 2^{ki} (1 + 2^k)^{-(k+1)i} -$$

$$- \frac{1}{2}\left(\frac{1 + 2^k}{2}\right)^{n-1} \sum_{i=0}^{[(n-1)/(k+1)]} (-1)^i \binom{n - 1 - k}{i} 2^{ki} \times$$

$$\times (1 + 2^k)^{-(k+1)i}, \quad \text{for } n \geqslant 1.$$

Part (a) of Corollary 3.4 was proposed by Swamy [27]. Parts (b) and (c) were shown in [20].

Theorem 3.3 [23]: Let X be a random variable distributed as $G_k(p)$. Then $P(X \leqslant n) = 0$ if $n < k$, and

$$P(X \leqslant n)$$

$$= 1 - \frac{p^{n+1}}{q} F_{n+2}^{(k)}(q/p)$$

$$= 1 - \frac{p^{n+1}}{q} \sum_{\substack{n_1, \ldots, n_k \geqslant \\ n_1 + 2n_2 + \cdots + kn_k = n+1}} \binom{n_1 + \cdots + n_k}{n_1, \ldots, n_k} \times$$

$$\times \left(\frac{q}{p}\right)^{n_1 + \cdots + n_k}, \quad \text{for } n \geqslant k.$$

The proof follows from Lemma 3.1 below and Theorem 3.2(a), both applied with $x = q/p$.

Lemma 3.1 [23]: Let $\{F_n^{(k)}(x)\}_{n=0}^{\infty}$ be the sequence of Fibonacci-type polynomials of order k. Then

$$\sum_{n=0}^{m} \frac{F_{n+1}^{(k)}(x)}{(1 + x)^{n+k}} = 1 - \frac{F_{m+k+2}^{(k)}(x)}{x(1 + x)^{m+k}}, \quad m \geqslant 0.$$

Now let L_n be a random variable that denotes the length of the longest run of successes in n ($\geqslant 1$) Bernoulli trials. The methodology of Philippou and Muwafi [19] and the Fibonacci-type polynomials of order k are instrumental in deriving the probability distribution of L_n, its probability generating function, and its factorial moments. The probability distribution of L_n will be employed in Section 4, along with Theorem 3.2, to give three new formulas for the reliability of a consecutive-k-out-of-n: F system, two of which are closed.

Theorem 3.4 [17]: Let $\{F_n^{(k)}(x)\}_{n=0}^{\infty}$ be the sequence of Fibonacci-type polynomials of order k, and denote by L_n the length of the longest success run in n ($\geqslant 1$) Bernoulli trials. Then

$$P(L_n \leqslant k) = \frac{p^{n+1}}{q} F_{n+2}^{(k+1)}(q/p), \quad 0 \leqslant k \leqslant n.$$

Remark 3.1: Theorem 3.4 also provides two closed formulas for the probability distribution of L_n, by means of Theorem 3.2.

Remark 3.2: A simple derivation of Theorem 3.4 may be obtained from Theorem 3.3 by noting that $(L_n \geqslant k + 1) = (N_{k+1} \leqslant n)$, where N_{k+1} is as in Section 2.

As a corollary of Theorem 3.4, we obtain the following formula,

$$P(L_n \geqslant k \mid p = 1/2) = 1 - F_{n+2}^{(k)}/2^n, \quad 1 \leqslant k \leqslant n, \quad (3.3)$$

which was the main result of McCarty [12]. Two enumeration theorems of Bollinger [3] also follow from Theorem 3.4 as simple corollaries. The same theorem gives the following, by means of some algebra.

Theorem 3.5 [17]: Let $\{F_n^{(k)}(x)\}_{n=0}^{\infty}$ and L_n be as in Theorem 3.4 and denote by $g_n(s)$ the probability generating function of L_n. Then

(a) $\quad g_n(s) = s^n - (t - 1)\dfrac{p^{n+1}}{q}\displaystyle\sum_{k=0}^{n-1} s^k F_{n+2}^{(k+1)}(q/p), \quad n \geqslant 1;$

(b) $\quad E(L_n^{(r)}) = n^{(r)} - r\dfrac{p^{n+1}}{q}\displaystyle\sum_{k=r-1}^{n-1} k^{(r-1)} F_{n+2}^{(k+1)}(q/p),$

$\quad 1 \leqslant r \leqslant n,$

where $m^{(0)} = 1$ and $m^{(r)} = m(m - 1) \cdots (m - r + 1)$, $r \geqslant 1$.

The probabilities $p_n = P(X = n)$ $(n \geqslant k)$ of the geometric distribution of order k may be easily calculated by means of the following theorem.

Theorem 3.6 [18]: Let X be a random variable distributed as $G_k(p)$ and set $p_n = P(X = n)$. Then

$$
p_n = \begin{cases}
p^k, & n = k, \\
qp^k, & k + 1 \leqslant n \leqslant 2k, \\
p_{n-1} - qp^k p_{n-1-k}, & n \geqslant 2k + 1.
\end{cases}
$$

Proof: The proof follows directly from Corollary 3.1(b) and the following simple proposition, upon setting $x = q/p$.

Proposition 3.1 [23]: Let $\{F_n^{(k)}(x)\}_{n=0}^{\infty}$ be the sequence of Fibonacci-type polynomials of order k. Then

$$
F_n^{(k)}(x) = \begin{cases}
x(1 + x)^{n-2}, & 2 \leqslant n \leqslant k + 1, \\
(1 + x)F_{n-1}^{(k)}(x) - xF_{n-1-k}^{(k)}(x), & n \geqslant k + 2.
\end{cases}
$$

Remark 3.3: The proof of [18] employs Theorem 3.3. An alternative proof of a variant of Theorem 3.6, based on first principles, is given independently by Aki, Kuboki, and Hirano [1].

4. RELIABILITY OF A CONSECUTIVE-k-OUT-OF-n : F SYSTEM

The reliability of a system may be increased without duplicating the system by using what reliability engineers call a consecutive-k-out-of-n : F system. Such a system consists of n ($\geqslant 1$) ordered components and fails if and only if at least k ($1 \leqslant k \leqslant n$) consecutive components fail. The consecutive-k-out-of-n : F system was first introduced by Chiang and Niu [6] in connection with telecommunication and oil pipeline systems, who obtained the following results.

Theorem 4.1 [6]: Assume that the components of the consecutive-k-out-of-n : F system are ordered linearly and function independently with probability p. Denote the reliability of the system by $R(p; k, n)$. Then

$$
\text{(a)} \quad R(p; 2, n) = \sum_{i=0}^{[(n+1)/2]} \binom{n - i + 1}{i} p^{n-i} q^i, \quad n \geqslant 2;
$$

(b) $R(p; k, n) = \sum\limits_{i=1}^{n-k+1} \sum\limits_{j=i+1}^{i+k-1} R(p; k, n - j)p^i q^{j-i} +$

$$+ p^{n-k+1}, \quad n \geqslant k,$$

where $R(p; k, n) = 1$ if $0 \leqslant n < k$ and 0 if $n < 0$;

(c) $(1 - q^k)^{n-k+1} \leqslant R(p; k, n) \leqslant (1 - q^k)^{[n/k]}, \quad n \geqslant k.$

It should be noted that the bounds in (c) above are sharp because the equalities both hold for $k = n$.

Derman, Lieberman, and Ross [7] derived another recursive formula for $R(p; k, n)$. They also obtained a recursive formula for the reliability $R_c(p; k, n)$ of a circular consecutive-k-out-of-n : F system.

Theorem 4.2 [7]: Let $R(p; k, n)$ be as in Theorem 4.1. Then

$$R(p; k, n) = \sum\limits_{i=0}^{n} N(i, n-i+1; k-1)p^{n-i}q^i, \quad n \geqslant k,$$

where $N(i, r; 1) = \binom{r}{i}$ if $0 \leqslant i \leqslant r$ and 0 if $i > r$, and

$$N(i, r; m) = \sum\limits_{j=0}^{r} \binom{r}{j} N(i - mj, r - j; m - 1), \quad m \geqslant 2.$$

Remark 4.1 [7]: Theorem 4.1(a) follows from Theorem 4.2 by setting $k = 2$.

Theorem 4.3 [7]: Assume that the components of the consecutive-k-out-of-n : F system are ordered circularly and function independently with probability p. Denote the reliability of the system by $R_c(p; k, n)$. Then

$$R_c(p; k, n) = p^2 \sum\limits_{i=0}^{k-1} (i + 1)q^i R(p; k, n - i - 2), \quad n \geqslant k.$$

Derman, Lieberman, and Ross [7] also considered the case of unequally reliable components and derived bounds for the reliability of both the linear and the circular consecutive-k-out-of-n : F system.

Bollinger [4] derived the following theorem.

Theorem 4.4 [4]: Let $R(p; k, n)$ be as in Theorem 4.1, and denote by $r_i^{(k)}$ the number of binary numbers of length n containing i ones, with at least k of these ones consecutive. Then

$$R(p; k, n) = 1 - \sum_{i=k}^{n} r_i^{(k)} p^{n-i} q^i, \quad n \geqslant k.$$

The coefficients $r_i^{(k)}$ are given by

$$r_i^{(k)} = \binom{n}{i} - C_k(n - i + 1, i), \quad k \leqslant i \leqslant n,$$

where $C_k(\ell, m)$ is the (ℓ, m) entry in the Pascal triangle of order k.

For more on the Pascal triangles of order k, we refer to [5]. We finally mention Shanthikumar [26], who derived an efficient recursive algorithm which computes the reliability of linear consecutive-k-out-of-n: F systems with unequal component reliabilities. For the case of equal component reliabilities his result reduces to the following theorem.

Theorem 4.5 [26]: Let $R(p; k, n)$ be as in Theorem 4.1. Then

(a) $R(p; k, n) = R(p; k, n-1) - R(p; k, n-1-k)pq^k$, $n \geqslant 2k + 1$, .

(b) $R(p; k, n) = 1 - q^k - (n-k)pq^k$, $k \leqslant n \leqslant 2k$.

It may be noted that Theorem 4.5 provides an excellent alternative to Theorems 4.1(b), 4.2, and 4.4.

We shall now derive three new formulas for the reliability $R(p; k, n)$, two of which are closed, by means of the distribution of the longest run and the expansions of the Fibonacci-type polynomials of order k. We shall also obtain the sharp upper bound of $R(p; k, n)$, given in Theorem 4.1(c), and offer a new proof of Theorem 4.5.

Theorem 4.6: Assume that the components of the consecutive-k-out-of-n: F system are ordered linearly and function independently with probability p. Let $\{F_n^{(k)}(x)\}_{n=0}^{\infty}$ by the Fibonacci-type polynomials of order k, and denote the reliability of the system by $R(p; k, n)$. Then

(a) $R(p;\ k,\ n) = \dfrac{q^{n+1}}{p}\ F^{(k)}_{n+2}(p/q),\ n \geqslant k;$

(b) $R(p;\ k,\ n) = \dfrac{q^{n+1}}{p} \displaystyle\sum_{\substack{n_1,\ \ldots,\ n_k\ \ni \\ n_1 + 2n_2 + \cdots + kn_k = n+1}} \binom{n_1 + \cdots + n_k}{n_1,\ \ldots,\ n_k} \times$

$$\times \left(\dfrac{p}{q}\right)^{n_1 + \cdots + n_k},\ n \geqslant k;$$

(c) $R(p;\ k,\ n) = \dfrac{1}{p} \displaystyle\sum_{i=0}^{[(n+1)/(k+1)]} (-1)^i \binom{n+1-ki}{i}(pq^k)^i -$

$$- \dfrac{q}{p} \displaystyle\sum_{i=0}^{[n/(k+1)]} (-1)^i \binom{n-ki}{i}(pq^k)^i,\ n \geqslant k.$$

Proof: Denote by $\tilde{L}_n$ the length of the longest failure run among the n components. Then, by the definition of $R(p;\ k,\ n)$,

$$R(p;\ k,\ n) = 1 - P(\tilde{L}_n \geqslant k).$$

By Theorem 3.4,

$$P(\tilde{L}_n \geqslant k) = 1 - P(\tilde{L}_n \leqslant k - 1)$$

$$= 1 - \dfrac{q^{n+1}}{p}\ F^{(k)}_{n+2}(p/q),\ n \geqslant k.$$

The last two relations establish part (a) of the theorem. Parts (b) and (c) follow from (a) by means of Theorem 3.2 applied with $x = p/q$.

Corollary 4.1: Let $R(p;\ k,\ n)$ be as in Theorem 4.6. Then

$$R(p;\ 2,\ n) = \displaystyle\sum_{i=0}^{[(n+1)/2]} \binom{n+1-i}{i} p^{n-i} q^i,\ n \geqslant 2.$$

Proof: The proof follows from Theorem 4.6(b) by setting $k = 2$.

Corollary 4.2: Let $R(p;\ k,\ n)$ be as in Theorem 4.6. Then

$$R(p;\ k,\ n) \leqslant (1 - q^k)^{[n/k]},\ n \geqslant k.$$

Equality is attained for $n = k$.

The proof is a direct consequence of Theorem 4.6 and the following lemma.

Lemma 4.1: Let $\{F_n^{(k)}(x)\}_{n=0}^{\infty}$ be the sequence of Fibonacci-type polynomials of order k. Then

$$F_{n+2}^{(k)}(q/p) \leqslant qp^{-(n+1)}(1 - p^k)^{[n/k]}, \ n \geqslant 0.$$

Equality is attained for $0 \leqslant n \leqslant k$.

Proof: Upon setting $x = q/p$, Proposition 3.1 gives

$$F_{n+2}^{(k)}(q/p) = qp^{-(n+1)}, \ 0 \leqslant n \leqslant k - 1;$$

$$F_{k+2}^{(k)}(q/p) = qp^{-(k+1)}(1 - p).$$

Also,

$$F_{n+2}^{(k)}(q/p)$$

$$= qp^{-(n+1)}P(L_n \leqslant k - 1), \ n \geqslant k, \text{ by Theorem 3.4,}$$

$$= qp^{-(n+1)}P(N_n^{(k)} = 0), \text{ by the definition of } N_n^{(k)},$$

$$= qp^{-(n+1)}\{1 - P(N_n^{(k)} \geqslant 1)\}$$

$$\leqslant qp^{-(n+1)}\{1 - \{1 - (1 - p^k)^{[n/k]}\}\}, \text{ by Proposition}$$
$$\hspace{9cm} 6.3 \text{ of } [2],$$

$$= qp^{-(n+1)}(1 - p^k)^{[n/k]}.$$

The last two relations establish the lemma.

We now proceed to give another proof of Theorem 4.5. First, we show the following lemma.

Lemma 4.2: Let $\{F_n^{(k)}(x)\}_{n=0}^{\infty}$ be the sequence of Fibonacci-type polynomials of order k. Then, for $k \leqslant n \leqslant 2k$,

$$F_{n+2}^{(k)}(p/q) = pq^{-(n+1)}\{1 - q^k\{1 + (n - k)p\}\}.$$

Proof: For $k + 1 \leqslant n \leqslant 2k$, it follows that $[(n+1)/(k+1)] = 1$ and $[n/(k+1)] = 1$. Then Theorem 3.2(b), applied with $x = p/q$, gives

$$F_{n+2}^{(k)}(p/q) = q^{-(n+1)} \sum_{i=0}^{1} (-1)^i \binom{n + 1 - ki}{i}(pq^k)^i -$$

$$- q^{-n} \sum_{i=0}^{1} (-1)^i \binom{n - ki}{i}(pq^k)^i$$

$$= pq^{-(n+1)}\{1 - q^k\{1 - (n - k)p\}\}, \quad k \leqslant n \leqslant 2k,$$

by means of some algebra. For $n = k$, the lemma follows from Proposition 3.1 upon setting $x = p/q$.

New Proof of Theorem 4.5: Upon setting $x = p/q$, Proposition 3.1 gives

$$F_{n+2}^{(k)}(p/q) = \frac{1}{q} F_{n+1}^{(k)}(p/q) - \frac{p}{q} F_{n+1-k}^{(k)}(p/q), \quad n \geqslant k,$$

from which (a) follows by means of Theorem 4.6(a).

   Part (b) is an immediate consequence of Theorem 4.6(a) and Lemma 4.2.

   We may also use Lemma 4.2 to give a partial solution to Problem 2.1(b).

Proposition 4.1: Let X be a random variable distributed as $B_k(n, p)$, and assume that

$$\sum_{x=0}^{[n/k]} P(X = x) = 1.$$

Then, for $k \leqslant n \leqslant 2k - 1$,

$$E(X) = p^k\{1 + (n - k)q\}$$

and

$$\sigma^2(X) = p^k\{1 + (n - k)q\} - p^{2k}\{1 + (n - k)q\}^2.$$

Proof: By means of Definition 2.7, Theorem 3.2(a), and Definition 3.2, we have

$$P(X = 0)$$

$$= p^n \sum_{i=0}^{k-1} \sum_{\substack{x_1, \ldots, x_k \ni \\ x_1 + 2x_2 + \cdots + kx_k = n-i}} \binom{x_1 + \cdots + x_k}{x_1, \ldots, x_k}\left(\frac{q}{p}\right)^{x_1 + \cdots + x_k}$$

$$= p^n \sum_{i=0}^{k-1} F_{n+1-i}^{(k)}(q/p) = \frac{p^{n+1}}{q} F_{n+2}^{(k)}(p/q).$$

By assumption,

$$\sum_{x=0}^{[n/k]} P(X = x) = 1,$$

so that

$$P(X = 1) = 1 - P(X = 0), \ k \leqslant n \leqslant 2k - 1,$$

$$= 1 - \frac{p^{n+1}}{q} F_{n+2}^{(k)}(q/p).$$

Therefore,

$$E(X^i) = 1 - \frac{p^{n+1}}{q} F_{n+2}^{(k)}(q/p), \ i \geqslant 1, \ k \leqslant n \leqslant 2k - 1,$$

$$= p^k\{1 + (n - k)q\}, \ \text{by Lemma 4.2,}$$

from which the proposition follows.

We end this paper by noting the following relation between the entries of the Pascal triangle of order k and the Fibonacci-type polynomials of order k. Its corollary may be compared with Theorem 3(ii) of [4].

Proposition 4.2: Let $\{F_n^{(k)}(x)\}_{n=0}^{\infty}$ be the sequence of Fibonacci-type polynomials of order k and let $C_k(\ell, m)$ be the (ℓ, m) entry in the Pascal triangle of order k. Then

$$F_{n+2}^{(k)}(p/q) = (p/q)^{n+1} \sum_{i=0}^{n} C_k(n - i + 1, \ i)(q/p)^i, \ n \geqslant k.$$

Proof: By means of the definition of $r_i^{(k)}$ and $C_k(\ell, m)$,

$$r_i^{(k)} = 0, \ 0 \leqslant i \leqslant k - 1,$$

and

$$C_k(n - i + 1, \ i) = \binom{n}{i}, \ 0 \leqslant i \leqslant k - 1.$$

Therefore, by Theorem 4.4,

$$R(p; \ k, \ n) = 1 - \sum_{i=k}^{n} r_i^{(k)} p^{n-i} q^i$$

$$= \sum_{i=0}^{n} \left\{\binom{n}{i} - r_i^{(k)}\right\} p^{n-i} q^i$$

$$= \sum_{i=0}^{n} C_k(n - i + 1, \ i) p^{n-i} q^i, \ n \geqslant k.$$

But

$$R(p; \ k, \ n) = \frac{q^{n+1}}{p} F_{n+2}^{(k)}(p/q), \ \text{by Theorem 4.6(a).}$$

The last two relations establish the proposition.

Corollary 4.3: Let $\{f_n^{(k)}\}_{n=0}^{\infty}$ be the Fibonacci numbers of order k, and let $C(\ell, m)$ be as in Proposition 4.2. Then

$$f_{n+2}^{(k)} = \sum_{i=0}^{n} C_k(n - i + 1, i), \ n \geqslant k.$$

REFERENCES

[1] Aki, S.; Kuboki, H.; and Hirano, K. 'On Discrete Distributions of Order k.' *Annals of the Institute of Statistical Mathematics* **36**, no. 3 (1984):431-440.

[2] Berman, S. M. *The Elements of Probability*. Reading, Mass.: Addison-Wesley, 1969.

[3] Bollinger, R. C. 'Fibonacci k-Sequences, Pascal-T Triangles and k-in-a-Row Problems.' *The Fibonacci Quarterly* **22**, no. 2 (1984):146-151.

[4] Bollinger, R. C. 'Direct Computation for Consecutive-k-out-of-n: F Systems.' *IEEE Transactions on Reliability* **R-31**, no. 5 (1982):444-446.

[5] Bollinger, R. C. 'A Note on Pascal-T Triangles, Multinomial Coefficients, and Pascal Pyramids.' *The Fibonacci Quarterly* **24** (1986), to appear.

[6] Chiang, D. T., and Niu, S. 'Reliability of Consecutive-k-out-of-n: F System.' *IEEE Transactions on Reliability* **R-30**, no. 1 (1981):87-89.

[7] Derman, C.; Lieberman, G. J.; and Ross, S. M. 'On the Consecutive-k-of-n: F System.' *IEEE Transactions on Reliability* **R-31**, no. 1 (1982):57-63.

[8] Feller, W. *An Introduction to Probability Theory and Its Applications*, Vol. I. 3rd ed. New York: Wiley, 1968.

[9] Hirano, K. 'Some Properties of the Distributions of Order k.' In *Fibonacci Numbers with Applications: Proceedings of the First International Conference on Fibonacci Numbers and Their Applications (Patras 1984)*. Edited by A. N. Philippou, A. F. Horadam, and G. E. Bergum. Dordrecht: D. Reidel, 1986, pp. 43-53.

[10] Hirano, K.; Kuboki, H.; Aki, S.; and Kuribayashi, A. *Figures of Probability Functions in Statistics II— Discrete Univariate Case*. Computer Science Monographs No. 20. Tokyo: The Institute of Statistical Mathematics, 1984.

[11] Hoggatt, V. E., Jr., and Bicknell, M. 'Generalized Fibonacci Polynomials.' *The Fibonacci Quarterly* 11, no. 5 (1973):457–465.

[12] McCarty, C. P. 'Coin Tossing and r-Bonacci Numbers.' In *A Collection of Manuscripts Related to the Fibonacci Sequence: 18th Anniversary Volume*. Edited by V. E. Hoggatt, Jr., and Marjorie Bicknell-Johnson. Santa Clara, Calif.: The Fibonacci Association, 1980, pp. 130–132.

[13] Philippou, A. N. 'A Note on the Fibonacci Sequence of Order k and the Multinomial Coefficients.' *The Fibonacci Quarterly* 21, no. 2 (1983):82–86.

[14] Philippou, A. N. 'Poisson and Compound Poisson Distributions of Order k and Some of Their Properties' (in Russian, English summary). *Zapiski Nauchnykh Seminarov Leningradskogo Otdeleniya Matematicheskogo Instituta im. V. A. Steklova AN SSSR* 130 (1983):175–180.

[15] Philippou, A. N. 'The Negative Binomial Distribution of Order k and Some of Its Properties.' *Biometrical Journal* 26, no. 7 (1984):789–794.

[16] Philippou, A. N., and Hadjichristos, J. H. 'A Note on the Poisson Distribution of Order k and a Result of Raikov.' *Institute of Mathematical Statistics Bulletin* 13, no. 6 (1984):368–369.

[17] Philippou, A. N., and Makri, F. S. 'Longest Success Runs and Fibonacci-Type Polynomials.' *The Fibonacci Quarterly* 23, no. 4 (1985):338–346.

[18] Philippou, A. N., and Makri, F. S. 'Successes, Runs and Longest Runs.' *Statistics and Probability Letters* 4, no. 1 (1986), in press.

[19] Philippou, A. N., and Muwafi, A. A. 'Waiting for the kth Consecutive Success and the Fibonacci Sequence of Order k.' *The Fibonacci Quarterly* 20, no. 1 (1982): 28–32.

[20] Philippou, A. N., and Philippou, G. N. 'The Pell Sequence of Order k, Multinomial Coefficients, and Probability.' *Bulletin of the Greek Mathematical Society* 22 (1984):74–84.

[21] Philippou, A. N.; Georghiou, C.; and Philippou, G. N. 'A Generalized Geometric Distribution and Some of Its Properties.' *Statistics and Probability Letters* 1, no. 4 (1983):171–175.

[22] Philippou, A. N.; Georghiou, C.; and Philippou, G. N. 'Fibonacci Polynomials of Order k, Multinomial

Expansions, and Probability.' *International Journal of Mathematics and Mathematical Sciences* 6, no. 3 (1983):545-550.

[23] Philippou, A. N.; Georghiou, C.; and Philippou, G. N. 'Fibonacci-Type Polynomials of Order *k* with Probability Applications.' *The Fibonacci Quarterly* 23, no. 2 (1985):100-105.

[24] Raikov, D. 'On the Decomposition of Gauss's and Poisson's Laws.' *Izvestia Akad. Nauk SSSR* 2 (1938): 91-124.

[25] Shane, H. D. 'A Fibonacci Probability Function.' *The Fibonacci Quarterly* 11, no. 6 (1973):517-522.

[26] Shanthikumar, G. J. 'Recursive Algorithm to Evaluate the Reliability of a Consecutive *k*-out-of-*n*: *F* System.' *IEEE Transactions on Reliability* R-31, no. 5 (1982): 442-443.

[27] Swamy, M. N. S. Problem B-74. *The Fibonacci Quarterly* 3, no. 3 (1965):236.

[28] Turner, S. J. 'Probability via the *n*th Order Fibonacci-*T* Sequence.' *The Fibonacci Quarterly* 17, no. 1 (1979):23-28.

[29] Uppuluri, V. R. R., and Patil, S. A. 'Waiting Times and Generalized Fibonacci Sequences.' *The Fibonacci Quarterly* 21, no. 4 (1983):342-349.

[30] Xekalaki, E.; Panaretos, J.; and Philippou, A.N. 'On Some Mixtures of Distributions of Order *k*. Submitted for publication (1985). A preliminary version of the paper was presented at the First International Conference on Fibonacci Numbers and Their Applications (Patras 1984).

G. N. Philippou and C. Georghiou

FIBONACCI-TYPE POLYNOMIALS AND PASCAL TRIANGLES OF ORDER k

1. INTRODUCTION

The Fibonacci-type polynomials of order k $(k \geqslant 2)$ have been introduced and studied by Philippou, Georghiou, & Philippou [1]. The Pascal triangle of order k was defined by Philippou [2]. In this paper a closed formula is given, expressing the Fibonacci-type polynomials in terms of the entries of the Pascal triangle of the same order.

2. FIBONACCI-TYPE POLYNOMIALS AND PASCAL TRIANGLES

We need the following definitions:

Definition 2.1: Let k be a fixed integer $(k \geqslant 2)$ and n be a nonnegative integer as specified. Then $\{F_n^{(k)}(x)\}_{n=0}^{\infty}$ is said to be the sequence of Fibonacci-type polynomials if $F_0^{(k)}(x) = 0$, $F_1^{(k)}(x) = 1$, and

$$F_n^{(k)}(x) = \begin{cases} x \sum\limits_{i=1}^{n} F_{n-i}^{(k)}(x), & 2 \leqslant n \leqslant k \\[2mm] x \sum\limits_{i=1}^{k} F_{n-i}^{(k)}(x), & n > k \end{cases} \tag{1}$$

Definition 2.2: The Pascal triangle of order k, denoted by T_k, is the rectangular array $A_{n,m}^{(k)}$ if $A_{0,0}^{(k)} = 1$, $A_{0,m}^{(k)} = 0$ for $m \geqslant 1$, and

$$A_{n,m}^{(k)} = \begin{cases} \sum\limits_{i=0}^{m} A_{n-1, m-i}^{(k)}, & 0 \leqslant m < k \text{ and } n \geqslant 1 \\[2mm] \sum\limits_{i=0}^{k-1} A_{n-1, m-i}^{(k)}, & m \geqslant k \text{ and } n \geqslant 1 \end{cases} \tag{2}$$

A. N. Philippou et al. (eds.), Fibonacci Numbers and Their Applications, 229–233.

Note that for $k = 2$ it follows directly from this definition that

$$A_{n,m}^{(2)} = \binom{n}{m},$$

i.e., the binomial coefficients.

The main result of this paper is the following theorem.

Theorem: Let $\{F_n^{(k)}(x)\}_{n=0}^{\infty}$ be the sequence of Fibonacci-type polynomials of order k and $A_{n,m}^{(k)}$ be the entries of the Pascal triangle of order k. Then

$$F_{n+1}^{(k)}(x) = \sum_{i=0}^{[n-n/k]} A_{n-i,\,i}^{(k)} x^{n-i}, \quad n \geqslant 0. \tag{3}$$

Proof: The proof is given in two steps.

Step 1. Note that (3) is true for $n = 0, 1$. Next, assume that (3) is true for $n = 0, 1, 2, \ldots, \ell$ ($\ell < k$). Then, for $n = \ell + 1$, we have

$$F_{\ell+2}^{(k)}(x) = x\left[F_{\ell+1}^{(k)}(x) + F_{\ell}^{(k)}(x) + \cdots + F_1^{(k)}(x)\right]$$

$$= x\left[A_{\ell,\,0}^{(k)} x^{\ell} + A_{\ell-1,\,1}^{(k)} x^{\ell-1} + \cdots + A_{1,\,\ell-1}^{(k)} x\right] +$$

$$+ x\left[A_{\ell-1,\,0}^{(k)} x^{\ell-1} + \cdots + A_{1,\,\ell-2}^{(k)} x\right] + \cdots +$$

$$+ x\left[A_{1,\,0}^{(k)} x\right] + x A_{0,\,0}^{(k)}$$

$$= A_{\ell+1,\,0}^{(k)} x^{\ell+1} + A_{\ell,\,1}^{(k)} x^{\ell} + \cdots + A_{1,\,\ell}^{(k)} x$$

$$= \sum_{i=0}^{\ell} A_{\ell+1-i,\,i}^{(k)} x^{\ell+1-i}$$

since

$$A_{\ell,\,0}^{(k)} = A_{\ell+1,\,0}^{(k)} = 1$$

$$A_{\ell-1,\,1}^{(k)} + A_{\ell-1,\,0}^{(k)} = A_{\ell,\,1}^{(k)}$$

$$A_{\ell-2,\,2}^{(k)} + A_{\ell-2,\,1}^{(k)} + A_{\ell-2,\,0}^{(k)} = A_{\ell-1,\,2}^{(k)} \qquad \text{etc.}$$

and

$$A_{1,\,\ell}^{(k)} = A_{0,\,0}^{(k)} = 1.$$

Step 2. We assume that (3) holds for $n = \ell - k + 1$, $\ell - k + 2, \ldots, \ell$ ($\ell \geqslant k$). We shall give the induction step for $\ell \equiv 0 \pmod{k}$. The other possible cases $\ell \equiv i \pmod{k}$, $i = 1, 2, \ldots, k - 1$, can be treated similarly.

Let $\ell = kr$ for some nonnegative integer r. Then we have

$$F_{\ell+2}^{(k)}(x) = x \sum_{i=0}^{kr-1} F_{kr+1-i}^{(k)}(x)$$

$$= x \left[\sum_{i=0}^{rk-r} A_{kr-i,\,i}^{(k)} x^{kr-i} + \sum_{i=0}^{kr-r-1} A_{kr-1-i,\,i}^{(k)} x^{kr-1-i} + \right.$$

$$\left. + \cdots + \sum_{i=0}^{kr-r-k+1} A_{kr+1-k-i,\,i}^{(k)} x^{kr+1-k-i} \right]$$

$$= x \left[A_{kr,\,0}^{(k)} x^{kr} + A_{kr-1,\,1}^{(k)} x^{kr-1} + A_{kr-2,\,2}^{(k)} x^{kr-2} + \right.$$

$$\left. + \cdots + A_{r,\,kr-r}^{(k)} x^{r} \right] + x \left[A_{kr-1,\,0}^{(k)} x^{kr-1} + \right.$$

$$\left. + A_{kr-2,\,1}^{(k)} x^{kr-2} + \cdots + A_{r,\,kr-r-1}^{(k)} x^{r} \right] +$$

$$+ \cdots + x \left[A_{kr+1-k,\,0}^{(k)} x^{kr+1-k} + \cdots + A_{r,\,kr-r-k+1}^{(k)} x^{r} \right]$$

$$= A_{kr+1,\,0}^{(k)} x^{kr+1} + A_{kr,\,1}^{(k)} x^{kr} + A_{kr-1,\,2}^{(k)} x^{kr-1} +$$

$$+ \cdots + A_{r+1,\,kr-r}^{(k)} x^{r+1}$$

$$= \sum_{i=0}^{kr-r} A_{kr+1-i,\,i}^{(k)} x^{kr+1-i}$$

$$= \sum_{i=0}^{\left[\ell+1-\frac{\ell+1}{k} \right]} A_{\ell+1-i,\,i}^{(k)} x^{\ell+1-i}$$

since

$$A_{kr,\,0}^{(k)} = A_{kr+1,\,0}^{(k)} = 1$$

$$A_{kr-1,\,1}^{(k)} + A_{kr-1,\,0}^{(k)} = A_{kr,\,1}^{(k)}$$

$$A_{kr-2,\,2}^{(k)} + A_{kr-2,\,1}^{(k)} + A_{kr-2,\,0}^{(k)} = A_{kr-1,\,2}^{(k)} \qquad \text{etc.}$$

and this completes the proof.

Remark: For $x = 1$, we have $F^{(k)}(1) = F^{(k)}$, the Fibonacci sequence of order k [3], and formula (3) reduces to

$$F^{(k)} = \sum_{i=0}^{[n-n/k]} A_{n-i,\,i}^{(k)}.$$

In particular, when $k = 2$, we have $F_n^{(2)}(1) = F_n$, where F_n is the usual Fibonacci sequence, and we get

$$F_n = \sum_{i=0}^{[n/2]} A_{n-i,\,i}^{(2)} = \sum_{i=0}^{[n/2]} \binom{n-i}{i},$$

which is a well-known formula.

We end this paper by showing that the entries $A_{n,m}^{(k)}$ of the Pascal triangle T_k are given by

$$A_{n,m}^{(k)} = \sum_{i=0}^{[m/k]} (-1)^i \binom{n}{i} \binom{m+n-1-ki}{n-1}, \tag{4}$$

thus enabling us to have a neat formula for the Fibonacci-type polynomials.

The generating function of the Fibonacci-type polynomials is (see [2]),

$$G_k(s;\,x) = \frac{s(1-s)}{1-s-sx(1-s^k)}$$

and using the binomial theorem we obtain the following expression for these polynomials (see [2]):

$$F_{n+1}^{(k)}(x) = \sum_{j=0}^{[n/(k+1)]} (-1)^j \binom{n-kj}{j} x^j (1+x)^{n-(k+1)j} -$$

$$- \sum_{j=0}^{[(n-1)/(k+1)]} (-1)^j \binom{n-1-kj}{j} x^j (1+x)^{n-1-(k+1)j}.$$

Now, by expanding $(1+x)^{n-(k+1)j}$ and $(1+x)^{n-1-(k+1)j}$, we get

$$F_{n+1}^{(k)}(x) = \sum_{j=0}^{\infty} \sum_{\ell=0}^{\infty} (-1)^j \binom{n-kj}{j} \binom{n-kj-j}{\ell} x^{j+\ell} -$$

$$- \sum_{j=0}^{\infty} \sum_{\ell=0}^{\infty} (-1)^j \binom{n-1-kj}{j} \binom{n-1-kj-j}{\ell} x^{j+\ell},$$

and by setting $j + \ell = n - i$ we see that the coefficient of x^{n-i} is

$$\sum_{j=0}^{\infty} (-1)^j \binom{n-kj}{j}\binom{n-kj-j}{n-j-i} -$$

$$- \sum_{j=0}^{\infty} (-1)^j \binom{n-1-kj}{j}\binom{n-1-kj-j}{n-j-i}$$

$$= \sum_{j=0}^{\infty} (-1)^j \binom{n-i}{j}\binom{n-1-kj}{n-1-i}$$

$$= \sum_{j=0}^{[i/k]} (-1)^j \binom{n-i}{j}\binom{n-1-kj}{n-1-i} = A_{n-i,i}^{(k)},$$

from which (4) follows.

REFERENCES

[1] Philippou, A. N.; Georghiou, C.; and Philippou, G. N. 'Fibonacci-Type Polynomials of Order K with Probability Applications.' *The Fibonacci Quarterly* 23, no. 2 (1985):100–105.
[2] Philippou, G. N. 'Fibonacci Polynomials of Order K and Probability Distributions of Order K.' Ph.D. dissertation, University of Patras, Patras, Greece, 1984.
[3] Philippou, A. N., and Muwafi, A. A. 'Waiting for the kth Consecutive Success and the Fibonacci Sequence of Order k.' *The Fibonacci Quarterly* 20, no. 1 (1982): 28–32.

Gerhard Rosenberger

A NOTE ON FIBONACCI AND RELATED NUMBERS IN THE THEORY OF 2 × 2 MATRICES

1. Let K be a commutative field, $a \in K^* = K\backslash\{0\}$ and $x \in K$. We define, recursively, the following sequence of elements of K.

$$T_0(a, x) = 0, \qquad T_1(a, x) = 1,$$

$$T_n(a, x) = xT_{n-1}(a, x) - aT_{n-2}(a, x), \qquad n \in N, \\ n \geqslant 2.$$

If $a = -1$, $x = 1$, and $n \geqslant 0$, then $T_n(a, x)$ is the nth Fibonacci number (in K).

If $a = 1$ and $n \geqslant 0$, then $T(a, x)$ is the nth Tschebyscheff polynomial in x (of the second kind).

By mathematical induction, we immediately obtain

$$T_n^2(a, x) - T_{n+1}(a, x)T_{n-1}(a, x) = a^{n-1} \qquad (1)$$

and

$$T_{mn}(a, x) = T_m(a^n, T_{n+1}(a, x) - \\ - aT_{n-1}(a, x))T_n(a, x), \qquad (2)$$

where $n, m \in \mathbb{N}$.

The purpose of this note is to discuss some properties of the $T_n(a, x)$.

2. The $T_n(a, x)$ occur in a natural way in the theory of the group $GL(2, K)$, the group of invertible 2 × 2 matrices with coefficients in K.

We use the notation $[A, B]$ for $ABA^{-1}B^{-1}$, the commutator of $A, B \in GL(2, K)$, and E for the identity matrix. Let tr A be the trace and det A be the determinant of $A \in GL(2, K)$. Now let $A, B \in GL(2, K)$ with $a = $ det A, $b = $ det B, $x = $ tr A, $y = $ tr B, and $z = $ tr AB. Then, for $n, m \geqslant 1$,

$$A^n = T_n(a, x)A - aT_{n-1}(a, x)E; \qquad (3)$$

235

A. N. Philippou et al. (eds.), Fibonacci Numbers and Their Applications, 235–240.
© 1986 by D. Reidel Publishing Company.

$$\text{tr } AB = \text{tr } A \text{ tr } B - \det B \text{ tr } AB^{-1}; \tag{4}$$

$$\text{tr}[A, B] = a^{-1}x^2 + b^{-1}y^2 + (ab)^{-1}z^2 -$$
$$- (ab)^{-1}xyz - 2; \tag{5}$$

$$\text{tr}[A^n, B^m] - 2 = (a^{n-1}b^{m-1})^{-1}T_n^2(a, x)T_m^2(b, y)$$
$$(\text{tr}[A, B] - 2). \tag{6}$$

On the other hand, we have the following.

Proposition 1: Let K be a commutative field of characteristic $p \neq 2$ or 0. Let $a, b, x, y, z \in K$, $a \neq 0$, $b \neq 0$. Let k_1, k_2 be defined by $k_1 = x^2 - 4a$, $k_2 = b(a^{-1}x^2 + b^{-1}y^2 + (ab)^{-1}z^2 - (ab)^{-1}xyz - 4)$, and $k_2 \neq 0$. Then there exist $A, B \in GL(2, K)$ with $\det A = a$, $\det B = b$, and $\text{tr } A = x$, $\text{tr } B = y$, $\text{tr } AB = z$ if and only if there exist $p, q, r \in K$, $q \neq 0$, with $k_1 p^2 + k_2 q^2 = r^2$.

Proof: The proof is analogous to the proof of Satz 6 in [4]; cf. also Proposition (1.5) in [2].

3.   Now let $M = \{((a, x), (b, y), (ab, z)) \,|\, a, b \in K^*;$ $x, y, z \in K\} \subset (K^* \times K)^3$. We consider the group $G(K)$ of transformations of M, which is generated by the following transformations:

$$\alpha: ((a, x), (b, y), (ab, z))$$
$$\mapsto ((b^{-1}, b^{-1}y), (ab, z), (a, x)),$$

$$\beta: ((a, x), (b, y), (ab, z))$$
$$\mapsto ((b, y), (a^{-1}, a^{-1}x), (a^{-1}b, a^{-1}xy - a^{-1}z)),$$

and

$$\delta: ((a, x), (b, y), (ab, z))$$
$$\mapsto ((b, y), (a, x), (ab, z)).$$

The generators α, β, and δ of $G(K)$ satisfy the following relations:

$$\alpha^6 = \alpha^3\beta^2 = \delta^2 = (\alpha\delta)^2 = (\beta\delta)^2 = 1.$$

In general, this set of relations is not a set of defining relations. But $G(K)$ is an epimorphic image of the general linear group $GL(2, \mathbb{Z})$, because $GL(2, \mathbb{Z})$ is generated by

$$R = \begin{pmatrix} 0 & -1 \\ 1 & 1 \end{pmatrix}, \quad S = \begin{pmatrix} 0 & -1 \\ 1 & 0 \end{pmatrix}, \quad T = \begin{pmatrix} 0 & 1 \\ 1 & 0 \end{pmatrix},$$

with defining relations

$$R^6 = R^3 S^2 = T^2 = (RT)^2 = (ST)^2 = E$$

(cf. [1]).

Now let $H(K)$ be the subgroup of $G(K)$ that is generated by α and β. $H(K)$ is an epimorphic image of the special linear group $SL(2, \mathbb{Z})$, because $SL(2, \mathbb{Z})$ is generated by

$$R = \begin{pmatrix} 0 & -1 \\ 1 & 1 \end{pmatrix} \quad \text{and} \quad S = \begin{pmatrix} 0 & -1 \\ 1 & 0 \end{pmatrix}$$

with defining relations

$$R^6 = R^3 S^2 = E.$$

Proposition 2: Let K be a commutative field of characteristic 0. Then $G(K) \cong GL(2, \mathbb{Z})$ and $H(K) \cong SL(2, \mathbb{Z})$.

Proof: Let $N = \{((1, x), (1, y), (1, z)) \mid x, y, z \in K\}$. N is a subset of M which is invariant under the operation of $G(K)$; that means $\varphi(N) \subset N$ for all $\varphi \in G(K)$.

Let $G(N, K)$ be the group that we obtain if we restrict the operation of $G(K)$ to N; $G(N, K)$ is an epimorphic image of $G(K)$.

It is well known that $G(N, K) \cong PGL(2, \mathbb{Z})$, the projective general linear group (cf. for instance [3] and [4]). Therefore, either $G(K) \cong GL(2, \mathbb{Z})$ or $G(K) \cong PGL(2, \mathbb{Z})$. Now $G(K)$ contains an element of order 6 but $PGL(2, \mathbb{Z})$ does not contain an element of order 6. Therefore, $G(K) \cong GL(2, \mathbb{Z})$, and also $H(K) \cong SL(2, \mathbb{Z})$.

4. From now on, let K be a finite field; that is, $K = GF(q)$, $q = p^m$ for a prime number p and a natural number m.

We write $G(q)$ instead of $G(GF(q))$. $G(q)$ is a finite group. By the map $R \mapsto \alpha$, $S \mapsto \beta$, we get an epimorphism f from $SL(2, \mathbb{Z})$ onto $G(q)$; and the kernel Γ^q of f is a normal subgroup of the $SL(2, \mathbb{Z})$ of finite index. From the arithmetical point of view, it is very interesting (and very difficult) to give a complete description of the kernel Γ^q of f.

A first step is the determination of the order of the element $\gamma = \alpha\beta$ in $G(q)$. The order of γ in $G(q)$ describes the conductor of Γ^q and gives the first statements to answer the question of whether or not Γ^q is a congruence subgroup (cf. [6] and [7]).

Now we have

$$\gamma((a, x), (b, y), (ab, z))$$
$$= ((a, x), (a^{-1}b, a^{-1}xy - a^{-1}z), (b, y)),$$

and by mathematical induction we obtain

$$\gamma^n((a, x), (b, y), (ab, z))$$
$$= ((a, x), (a^{-n}b, a^{-n}yT_{n+1}(a, x) - a^{-n}zT_n(a, x)),$$
$$(a^{1-n}b, a^{1-n}yT_n(a, x) - a^{1-n}zT_{n-1}(a, x))),$$
$$n \in \mathbb{N}. \tag{7}$$

If $a^n = 1$, $T_n(a, x) = 0$, and $T_{n+1}(a, x) = 1$, then $-aT_{n-1}(a, x) = 1$ because of (1).

Therefore, we obtain

Proposition 3: Let $n \in \mathbb{N}$. Then $\gamma^n = 1$ if and only if $a^n = 1$, $T_n(a, x) = 0$, and $T_{n+1}(a, x) = 1$ for all $a \in GF(q)\setminus\{0\}$ and all $x \in GF(q)$.

Proposition 4: The order $O(\gamma)$ of γ is $p(q^2 - 1)$.

Proof: Because $G(q)$ is finite, γ has finite order $O(\gamma) \geqslant 2$. Let

$$r = \begin{cases} 2(q^2 - 1) & \text{if } p = 2, q = 2^m \\ \tfrac{1}{2}p(q^2 - 1) & \text{if } p \geqslant 3, q = p^m. \end{cases}$$

Let $A \in GL(2, GF(q))$ with $\det A = a$ and $\operatorname{tr} A = x$, and let

$$C = \begin{pmatrix} a^{-1} & 0 \\ 0 & a^{-1} \end{pmatrix} \in GL(2, GF(q)) \text{ and } B = CA^2;$$

C lies in the center of $GL(2, GF(q))$.

Because $\det B = 1$, the order of B divides r (cf. [1]).

Now $A^2 = C^{-1}B$; this means that the order of A divides $2r$.

Therefore, $O(\gamma)$ divides $2r$ by (2) and (3). On the other hand, r divides $O(\gamma)$ because

$$N = \{((1, x), (1, y), (1, z))|x, y, z \in GF(q)\} \subset M$$

(cf. [5]).

Thus, either $O(\gamma) = r$ or $O(\gamma) = 2r$. We shall show that $O(\gamma) = r$ if $p = 2$ and $O(\gamma) = 2r$ if $p \geqslant 3$.

(i)   Let $p = 2$. Let $L = GF(2^k)$, $k \in \mathbb{N}$. If $A \in GL(2, L)$ with $A^2 = E$, then det $A = 1$; if $A \in GL(2, L)$ with det $A^2 = 1$, then det $A = 1$. Therefore, $O(\gamma) = r$ (cf. [5]).

(ii)   Let $q \equiv 1 \pmod 4$. If $a \in GF(q)\backslash\{0\}$ is an element of order $q - 1$ in $GF(q)$, then the matrix

$$A = \begin{pmatrix} 0 & 1 \\ a & 0 \end{pmatrix}$$

has the order $2(q - 1)$. Therefore, $O(\gamma) = 2r$ because of (3).

(iii) Let $q \equiv 3 \pmod 4$. Let $a \in GF(q^2)\backslash\{0\}$ be an element of order $q^2 - 1$. Then $b = a^{(q-1)/2}$ has the order $2(q + 1)$. Let $x = b - b^{-1}$. Because $b^q = -b^{-1}$, we obtain $x \in GF(q)$. The matrix

$$A = \begin{pmatrix} b & 0 \\ 0 & -b^{-1} \end{pmatrix}$$

has tr $A = x \in GF(q)$ and det $A = -1 \in GF(q)$. The order of A is $2(q + 1)$. Therefore, $O(\gamma) = 2r$ because of (3).

REFERENCES

[1] Coxeter, H. S. M., and Moser, W. O. J. 'Generators and
     Relations for Discrete Groups.' In *Ergebnisse der
     Mathematik und ihrer Grenzgebiete*, 14. Berlin: Heidel-
     berg; New York: Springer, 1972.

[2] Kern-Isberner, G., and Rosenberger, G. 'Über Diskre-
     theitsbedingungen und die diophantische Gleichung
     $ax^2 + by^2 + cz^2 = dxyz$.' *Archiv der Mathematik* 34
     (1980):481-493.

[3] Rosenberger, G. 'Fuchssche Gruppen, die freies Produkt
     zweier zyklischer Gruppen sind, und die Gleichung
     $x^2 + y^2 + z^2 = xyz$.' Dissertation, Hamburg, 1972. In
     part: *Math. Annalen* 199 (1972):213-227.

[4] Rosenberger, G. 'Zu Fragen der Analysis im Zusammen-
     hang mit der Gleichung $x_1^2 + \cdots + x_n^2 - ax_1 \ldots x_n = b$.'
     Monatshefte fur Mathematik 85 (1977):211-233.

[5] Rosenberger, G. 'Über Tschebyscheff-Polynome, Nicht-
     Kongruenzuntergruppen der Modulgruppe und Fibonacci-
     Zahlen.' *Math. Annalen* 246 (1980):193-203.

[6] Schoeneberg, B. 'Elliptic Modular Functions.' In
     Grundlehren der mathematischen Wissenschaften, 203.
     Berlin: Heidelberg; New York: Springer, 1974.

[7] Wohlfahrt, K. 'An Extension of F. Klein's Level Con-
     cept.' *Illinois Journal of Mathematics* 8 (1964):529-
     535.

A. Rotkiewicz

PROBLEMS ON FIBONACCI NUMBERS AND THEIR GENERALIZATIONS

INTRODUCTION

Let $L_n = (\alpha^n - \beta^n)/(\alpha - \beta)$, where α and β are different roots of the trinomial $x^2 - Lx + Q$ ($L > 0$ and Q are rational integers and $D = L^2 - 4Q > 0$). Let $2 \nmid nm$, $(m, n) = 1$, $(L, Q) = 1$, and (n/m) is the Jacobi symbol. Then

(a) If $2 \mid L$, $Q \equiv 1 \pmod 4$, then $\left(\dfrac{L_n}{L_m}\right) = \left(\dfrac{n}{m}\right)$.

(b) If $2 \mid L$, $Q \equiv -1 \pmod 4$ or $4 \mid Q$, then $\left(\dfrac{L_n}{L_m}\right) = 1$.

(c) If $2 \| Q$, then $\left(\dfrac{L_n}{L_m}\right) = (-1)^\lambda$, where λ is the number

of terms in the formula

$$\frac{n}{m} = k_1 + \frac{1\rfloor}{\lfloor k_2} + \cdots + \frac{1\rfloor}{\lfloor k_\lambda},$$

which represents the expansion of the rational number n/m into a single continued fraction with $k_\lambda > 1$.

The Jacobi symbol (u_n/u_m), where u_n is the nth Fibonacci number, is also considered. A composite number n is called a Fibonacci pseudoprime if $u_{n - (5/n)} \equiv 0 \pmod n$. Several problems on these are discussed.

   Chao Ko [2] and Terjanian [22] prove the insolubility of some diophantine equations. The common idea of their proofs is to evaluate some Jacobi symbols in two ways: (1) without any assumption about the equation and (2) with the assumption that the equation has a solution.

   According to the theorem of Terjanian, the equation $x^{2n} + y^{2n} = z^{2n}$, where p is an odd prime, has no integer solutions if $2p \nmid x$ and $2p \nmid y$. To prove this theorem, Terjanian calculates the symbol

241

A. N. Philippou et al. (eds.), Fibonacci Numbers and Their Applications, 241–255.

$$\left(\frac{A_m(x, y)}{A_n(x, y)}\right),$$

where $2 \nmid mn$, $A_i(x, y) = (x^i - y^i)/(x - y)$, $4 \mid x - y$.

Chao Ko [2], in order to prove that the Catalan equation $x^2 - 1 = y^p$, where p is an odd prime > 3, has no integer solutions, calculates the symbol

$$\left(\frac{Q_p(y)}{Q_q(y)}\right), \quad \text{where } Q_n = \frac{y^n - (-1)^n}{y - (-1)}.$$

A similar method is used in the proof of the theorem that, in the Fibonacci series $u_1 = 1$, $u_2 = 1$, $u_{n+2} = u_n + u_{n+1}$, only the first, second, and twelfth terms are squares (see [8], [24]).

In [20] I applied similar ideas to the diophantine equations $P_n = \square$, $P_n = p\square$ (p is an odd prime), where P_n is the Lehmer number.

$$P_n(\alpha, \beta) = \begin{cases} (\alpha^n - \beta^n)/(\alpha - \beta) & \text{for } n \text{ odd,} \\ (\alpha^n - \beta^n)/(\alpha^2 - \beta^2) & \text{for } n \text{ even,} \end{cases}$$

α and β are roots of the trinomial $z^2 - \sqrt{L}z + Q$, $L > 0$ and Q are rational integers.

In that same paper, I found the formula for (P_n/P_m) in the case $2 \mid LQ$. This formula implies the following theorem.

Theorem 1: Let $L_n = (\alpha^n - \beta^n)/(\alpha - \beta)$, where α and β are different roots of the trinomial $x^2 - Lx + Q$ ($L > 0$ and Q are rational integers and $D = L^2 - 4Q > 0$). Let $2 \nmid mn$, $(m, n) = 1$, $(L, Q) = 1$. Then

(a) If $2 \mid L$, $Q \equiv 1 \pmod 4$, then $(L_n/L_m) = (n/m)$.

(b) If $2 \mid L$, $Q \equiv -1 \pmod 4$ or $4 \mid Q$, $L \equiv \pm 1 \pmod 4$, then $(L_n/L_m) = 1$.

(c) If $2 \| Q$, $L \equiv \pm 1 \pmod 4$, then $(L_n/L_m) = (-1)^\lambda$, where λ is the number of terms in the formula
$$\frac{n}{m} = k_1 + \frac{1}{\lfloor k_2} + \cdots + \frac{1}{\lfloor k_\lambda},$$ which represents the expansion of the rational number n/m into a simple continued fraction with $k_\lambda > 1$.

From the above theorem follows the impossibility of the diophantine equation $x^p + y^p = z^2$, which is established under the conditions $(x, y) = 1$, and $p|z$, $2 \nmid z$, or $p \nmid z$, $2|z$ (p prime > 3) (see [28]), thereby constituting a generalization of a recent result of Terjanian. It is worth noting that Nagell [9] proved the impossibility of the equation $x^5 + y^5 = z^2$ under the assumption $(x, y) = 1$, $5|z$.

Now, let u_n denote the nth term of the Fibonacci sequence

$$1, 1, 2, 3, 5, 8, 13, \ldots \quad (u_{n+2} = u_{n+1} + u_n)$$

and v_n denote the nth term of its associated sequence defined by $v_1 = 1$, $v_2 = 3$, $v_n = v_{n-1} + v_{n-2}$ $(1, 3, 4, 7, 11, \ldots)$.

Now we shall consider Jacobi's symbol (u_n/u_m), where n and m are coprime positive integers and $3 \nmid nm$.

Let n and m be coprime positive integers and $3 \nmid nm$. We can write the following sequence of equalities:

$$
\begin{aligned}
n &= k_1 m &&+ \varepsilon_1 r_1, & 0 &< r_1 < m \\
m &= k_2 r_1 &&+ \varepsilon_2 r_2, & 0 &< r_2 < r_1 \\
r_1 &= k_3 r_2 &&+ \varepsilon_3 r_3, & 0 &< r_3 < r_2 \\
&\ \vdots && \ \vdots && \ \vdots \\
r_{\ell-3} &= k_{\ell-1} r_{\ell-2} + \varepsilon_{\ell-1} r_{\ell-1}, & 0 &< r_{\ell-1} < r_{\ell-2} \\
r_{\ell-2} &= k_\ell r_{\ell-1} &&+ \varepsilon_\ell r_\ell, & r_\ell &= 1
\end{aligned}
\tag{1}
$$

$$\varepsilon_i = \pm 1, \ 3 \nmid r_i \text{ for } i = 1, 2, \ldots, \ell$$

Lemma 1: Let $(n, m) = 1$, $3 \nmid mm$, $n = km + \varepsilon r$, $\varepsilon = \pm 1$, $3 \nmid r$, then

$$\left(\frac{u_n}{u_m}\right) = \left(\frac{u_{m-1}}{u_m}\right)^k \left(\frac{u_r}{u_m}\right). \tag{2}$$

Proof of Lemma 1: First consider the case $n = km + r$, where $0 < r < m$. We have

$$2U_{km+r} = V_r U_{km} + V_{km} U_r. \tag{3}$$

But $V_n = 2U_{n-1} + U_n$, and from (3) it follows that

$$2U_n = V_r U_{km} + (2U_{km-1} + U_{km})U_r,$$

and since $U_m \mid U_{km}$, we have

$$\left(\frac{2}{U_m}\right)\left(\frac{U_n}{U_m}\right) = \left(\frac{2}{U_m}\right)\left(\frac{U_{km-1}}{U_m}\right)\left(\frac{U_r}{U_m}\right),$$

hence

$$\left(\frac{U_n}{U_m}\right) = \left(\frac{U_{km-1}}{U_m}\right)\left(\frac{U_r}{U_m}\right). \tag{4}$$

We now have

$$\left(\frac{U_{km-1}}{U_m}\right) = \left(\frac{U_{m-1}}{U_m}\right)^k. \tag{5}$$

This is true for $k = 1$. Suppose that (5) holds for a positive integer k. From the formula ([5], p. 12),

$$U_{m+n} = U_m U_{n-1} + U_{m+1} U_n, \tag{6}$$

it follows that

$$U_{(k+1)m-1} = U_{(km-1)+m} = U_{km-1} U_{m-1} + U_{km} U_m;$$

hence,

$$\left(\frac{U_{(k+1)m-1}}{U_m}\right) = \left(\frac{U_{km-1}}{U_m}\right)\left(\frac{U_{m-1}}{U_m}\right)$$

$$= \left(\frac{U_{m-1}}{U_m}\right)^k\left(\frac{U_{m-1}}{U_m}\right) = \left(\frac{U_{m-1}}{U_m}\right)^{k+1},$$

and (5) follows by induction.

Now we consider the case $\varepsilon = -1$. Then $n = km - r$, where $0 < r < m$, $3 \nmid r$. We have

$$2(-1)^r U_{km-r} = V_r U_{km} + V_{km}(-U_r);$$

hence,

$$2(-1)^r U_n = V_r U_{km} + (2U_{km-1} + U_{km})(-U_r). \tag{7}$$

From (7), it follows that

$$\left(\frac{2}{U_m}\right)\left(\frac{(-1)^r}{U_m}\right)\left(\frac{U_n}{U_m}\right) = \left(\frac{2}{U_m}\right)\left(\frac{U_{km-1}}{U_m}\right)\left(\frac{-1}{U_m}\right)\left(\frac{U_r}{U_m}\right);$$

hence,

$$\left(\frac{(-1)^r}{U_m}\right)\left(\frac{U_n}{U_m}\right) = \left(\frac{U_{km-1}}{U_m}\right)\left(\frac{-1}{U_m}\right)\left(\frac{U_r}{U_m}\right). \tag{8}$$

Since the residues of U_n mod 4 form a periodic sequence of order 6, given by $0, 1, 1, 2, 3, 1$, we have $U_m \equiv 1 \pmod 4$ for $m \not\equiv 4 \pmod 6$, $3 \nmid m$ and

$$\left(\frac{(-1)^r}{U_m}\right) = \left(\frac{-1}{U_m}\right) = 1, \text{ for } m \not\equiv 4 \pmod 6, \ 3 \nmid m.$$

If $m \equiv 4 \pmod 6$, then from $n = km - r$, $(n, m) = 1$, it follows that $2 \nmid r$ and also

$$\left(\frac{(-1)^r}{U_m}\right) = \left(\frac{-1}{U_m}\right).$$

Thus, from (5) and (8), it follows that

$$\left(\frac{U_n}{U_m}\right) = \left(\frac{U_{km-1}}{U_m}\right)\left(\frac{U_r}{U_m}\right) = \left(\frac{U_{m-1}}{U_m}\right)^k\left(\frac{U_r}{U_m}\right).$$

This completes the proof of Lemma 1.

Lemma 2: We have

$$\left(\frac{U_{m-1}}{U_m}\right) = (-1)^\ell \text{ for } m = 6\ell - 2, \ 6\ell - 1, \ 6\ell + 1, \ 6\ell + 2.$$
$$(9)$$

Proof of Lemma 2: First we note that it is enough to prove our Lemma for a positive integer ℓ. Indeed, if $\ell = -\ell_1$, where $\ell_1 \geqslant 1$, then from formula $U_{-n} = (-1)^{n-1}U_n$ and formula (9) for $\ell \geqslant 0$, it follows that

$$\left(\frac{U_{6\ell-3}}{U_{6\ell-2}}\right) = \left(\frac{U_{-6\ell_1-3}}{U_{-6\ell_1-2}}\right) = \left(\frac{U_{6\ell_1+3}}{-U_{6\ell_1+2}}\right) = \left(\frac{U_{6\ell_1+3}}{U_{6\ell_1+2}}\right)$$
$$= (-1)^{\ell_1} = (-1)^\ell.$$

Similarly, we check our formula in other cases.

Let $\ell \geqslant 0$, then we have

$$\left(\frac{U_{6\ell+1}}{U_{6\ell+2}}\right) = \left(\frac{U_{6\ell+2}}{U_{6\ell+1}}\right) = \left(\frac{U_{6\ell}}{U_{6\ell+1}}\right)$$

and

$$\left(\frac{U_{6\ell+1}}{U_{6\ell+2}}\right) = \left(\frac{U_{6\ell+2}}{U_{6\ell+1}}\right) = \left(\frac{2U_{6\ell+1} - U_{6\ell-1}}{U_{6\ell+1}}\right) = \left(\frac{U_{6\ell-1}}{U_{6\ell+1}}\right)$$

(continued)

$$= \left(\frac{U_{6\ell+1}}{U_{6\ell-1}}\right) = \left(\frac{2U_{6\ell-1} + U_{6\ell-2}}{U_{6\ell-1}}\right) = \left(\frac{U_{6\ell-2}}{U_{6\ell-1}}\right)$$

$$= \left(\frac{U_{6\ell-1}}{U_{6\ell-2}}\right) = \left(\frac{U_{6\ell-3}}{U_{6\ell-2}}\right);$$

hence

$$\left(\frac{U_{6\ell+1}}{U_{6\ell+2}}\right) = \left(\frac{U_{6\ell}}{U_{6\ell+1}}\right) = \left(\frac{U_{6\ell-2}}{U_{6\ell-1}}\right) = \left(\frac{U_{6\ell-3}}{U_{6\ell-2}}\right).$$

Thus, it is enough to prove that

$$\left(\frac{U_{6\ell+1}}{U_{6\ell+2}}\right) = (-1)^\ell \text{ for } \ell \geqslant 0.$$

For $\ell = 0$, we have $(U_1/U_2) = (1/1) = 1 = (-1)^0$. Suppose that our formula holds for a positive integer ℓ. We have

$$\left(\frac{U_{6(\ell+1)+1}}{U_{6(\ell+1)+2}}\right) = \left(\frac{U_{6\ell+7}}{U_{6\ell+8}}\right) = \left(\frac{2U_{6\ell+7} - U_{6\ell+5}}{U_{6\ell+7}}\right) = \left(\frac{U_{6\ell+5}}{U_{6\ell+7}}\right)$$

$$= \left(\frac{U_{6\ell+7}}{U_{6\ell+5}}\right) = \left(\frac{2U_{6\ell+5} + U_{6\ell+4}}{U_{6\ell+5}}\right) = \left(\frac{U_{6\ell+5}}{U_{6\ell+4}}\right)$$

$$= \left(\frac{2U_{6\ell+4} - U_{6\ell+2}}{U_{6\ell+4}}\right) = -\left(\frac{U_{6\ell+2}}{U_{6\ell+4}}\right) = -\left(\frac{U_{6\ell+4}}{U_{6\ell+2}}\right)$$

$$= -\left(\frac{2U_{6\ell+2} + U_{6\ell+1}}{U_{6\ell+2}}\right) = -\left(\frac{U_{6\ell+1}}{U_{6\ell+2}}\right)$$

$$= -(-1)^\ell = (-1)^{\ell+1}$$

and our formula is proved by induction.

Theorem 2: Let $(n, m) = 1$, $3 \nmid nm$, $m = 6\ell_1 + r$, $r_i = 6\ell_{\ell+1} + r$ ($r = \pm 1, \pm 2$; $i = 1, 2, \ldots, \ell - 1$), then

$$\left(\frac{U_n}{U_m}\right) = \left(\frac{U_{m-1}}{U_m}\right)^{k_1} \left(\frac{U_{r_1-1}}{U_{r_1}}\right)^{k_2} \cdots \left(\frac{U_{r_{\ell-1}-1}}{U_{r_{\ell-1}}}\right)^{k_\ell}$$

$$= (-1)^{\ell_1 k_1 + \ell_2 k_2 + \cdots + \ell_1 k_\ell},$$

where r_i, k_i ($i = 1, 2, \ldots, \ell$) are numbers defined by conditions (1).

Proof of Theorem 2: Since $(m, r_1) = 1$, $3 \nmid mr_1$, one of the numbers U_{r_1}, U_m is $\equiv 1 \pmod 4$ and

$$\left(\frac{U_{r_1}}{U_m}\right) = \left(\frac{U_m}{U_{r_1}}\right).$$

Similarly, since $3 \nmid r_{i-1}r_i$, $(r_{i-1}, r_i) = 1$, one of the numbers $U_{r_{i-1}}$, U_{r_i} is $\equiv 1 \pmod 4$ ($i = 1, 2, \ldots, \ell$) and

$$\left(\frac{U_{r_i}}{U_{r_{i-1}}}\right) = \left(\frac{U_{r_{i-1}}}{U_{r_i}}\right) \quad (i = 1, 2, \ldots, \ell),$$

and by Lemma 1 and Lemma 2 we have

$$\left(\frac{U_n}{U_m}\right) = \left(\frac{U_{m-1}}{U_m}\right)^{k_1}\left(\frac{U_{r_1}}{U_m}\right) = \left(\frac{U_{m-1}}{U_m}\right)^{k_2}\left(\frac{U_m}{U_{r_1}}\right)$$

$$= \left(\frac{U_{m-1}}{U_m}\right)^{k_1}\left(\frac{U_{r_1-1}}{U_{r_1}}\right)^{k_2}\left(\frac{U_{r_2}}{U_{r_1}}\right) = \left(\frac{U_{m-1}}{U_m}\right)^{k_1}\left(\frac{U_{r-1}}{U_{r_1}}\right)^{k_2}\left(\frac{U_{r_1}}{U_{r_2}}\right)$$

$$= \left(\frac{U_{m-1}}{U_m}\right)^{k_1}\left(\frac{U_{r_1-1}}{U_{r_1}}\right)^{k_2}\left(\frac{U_{r_2-1}}{U_{r_2}}\right)^{k_3}\left(\frac{U_{r_3}}{U_{r_2}}\right)$$

$$= \left(\frac{U_{m-1}}{U_m}\right)^{k_1}\left(\frac{U_{r_1-1}}{U_{r_1}}\right)^{k_2}\left(\frac{U_{r_2-1}}{U_{r_2}}\right)^{k_3} \cdots \left(\frac{U_{r_{\ell-1}-1}}{U_{r_{\ell-1}}}\right)^{k_\ell}\left(\frac{U_{r_\ell}}{U_{r_{\ell-1}}}\right)$$

$$= \left(\frac{U_{m-1}}{U_m}\right)^{k_1}\left(\frac{U_{r_1-1}}{U_{r_1}}\right)^{k_2}\left(\frac{U_{r_2-1}}{U_{r_2}}\right)^{k_3} \cdots \left(\frac{U_{r_{\ell-1}-1}}{U_{r_{\ell-1}}}\right)^{k_\ell}$$

$$= (-1)^{k_1\ell_1 + k_2\ell_2 + \cdots + k_\ell\ell_\ell}.$$

This completes the proof of Theorem 2.

Example 1: We shall calculate the symbol

$$\left(\frac{U_{37}}{U_{29}}\right).$$

We have

$$37 = 1 \cdot 29 + 8$$
$$29 = 3 \cdot 8 + 5$$
$$8 = 2 \cdot 5 - 2$$
$$5 = 2 \cdot 2 + 1$$

and by Theorem 2, we obtain

$$\left(\frac{U_{37}}{U_{29}}\right) = \left(\frac{U_{28}}{U_{29}}\right)^1 \left(\frac{U_7}{U_8}\right)^1 \left(\frac{U_4}{U_5}\right)^2 \left(\frac{U_1}{U_2}\right)^2 = \left(\frac{U_{28}}{U_{29}}\right)\left(\frac{U_7}{U_8}\right)$$

$$= \left(\frac{U_{6\cdot5-2}}{U_{6\cdot5-1}}\right)\left(\frac{U_{6\cdot1+1}}{U_{6\cdot1+2}}\right) = (-1)^5 (-1)^1 = 1.$$

Example 2: We shall calculate the symbol

$$\left(\frac{U_{13}}{U_{11}}\right).$$

We have

$$19 = 1 \cdot 11 + 8$$
$$11 = 2 \cdot 8 - 5$$
$$8 = 2 \cdot 5 - 2$$
$$5 = 2 \cdot 2 - 1$$

and by Theorem 2, we obtain

$$\left(\frac{U_{19}}{U_{18}}\right) = \left(\frac{U_{10}}{U_{11}}\right)^1 \left(\frac{U_7}{U_8}\right)^2 \left(\frac{U_4}{U_5}\right)^2 \left(\frac{U_1}{U_2}\right)^2 = \left(\frac{U_{10}}{U_{11}}\right)$$

$$= \left(\frac{U_{2\cdot6-2}}{U_{2\cdot6-1}}\right) = (-1)^2 = 1.$$

FIBONACCI PSEUDOPRIMES

A composite n is called a pseudoprime if $n \mid 2^n - 2$. Chinese mathematicians claimed twenty-five centuries ago that such numbers do not exist. Leibniz in September of 1680 stated incorrectly that pseudoprimes do not exist. The first proof of the existence of infinitely many pseudoprimes was given by M. Cipolla in 1904 (see [3]). Many papers on pseudo-primes have appeared (see my book of 1972 [16] and the long paper by Pomerance, Selfridge, and Wagstaff, 1980 [11]).

Let $U_n = (\alpha^n - \beta^n)/(\alpha - \beta)$ denote the nth Fibonacci number and $V_n = \alpha^n + \beta^n$, where α and β are roots of the trinomial $x^2 - x - 1$.

If p is prime, then

(a) $U_{p-(5/p)} \equiv 0 \pmod{p}$

(b) $U_p \equiv (5/p) \pmod{p}$

(c) $V_p \equiv 1 \pmod{p}$

Composite numbers p which satisfy the relations (a)-(c) are called Fibonacci psuedoprimes of type (a), (b), or (c).

H. J. A. Duparc (1955) [4] proved that there exist infinitely many Fibonacci pseudoprimes of the above types. He tabulated all pseudoprimes of type (c) that are < 555200. Duparc found that U_{2p}, where p is a prime $\geqslant 7$, is a Fibonacci pseudoprime of type (a). The same is true for pq, where $p \equiv 17 \pmod{60}$ and $q = p + 2$ are both prime. Also, E. Lehmer (1964) [6] showed that there are infinitely many Fibonacci pseudoprimes of type (a). Parberry [10] proved four theorems concerning congruences (a) and (b), and Yorinaga (1976) [25], [26], and [27] has studied the Fibonacci pseudoprimes and gives a table of the composite numbers $p < 707000$ that satisfy (c).

An odd composite number n will be called a Euler Fibonacci pseudoprime (efpsp) if

$$U_{(n-(5/n))/2} \equiv 0 \pmod{n} \text{ if } (-1/n) = 1, \tag{11}$$

or

$$V_{(n-(5/n))/2} \equiv 0 \pmod{n} \text{ if } (-1/n) = -1. \tag{12}$$

An odd composite number n is a strong Fibonacci pseudoprime (sfpsp) with $n - (5/n) = d \cdot 2^s$, d odd; thus, we have either

$$U_d \equiv 0 \pmod{n}, \tag{13}$$

or

$$V_{d \cdot 2^r} \equiv 0 \pmod{n}, \tag{14}$$

for some r with $0 \leqslant r < s$.

Every prime n satisfies the above conditions. These definitions are particular cases of Euler Lucas pseudoprimes and strong Lucas pseudoprimes. Let $U_n = (\alpha^n - \beta^n)/(\alpha - \beta)$, where α and $\beta \neq \alpha$ are roots of the trinomial $x^2 - Px + Q$, where $P > 0$ and Q are rational integers, $D = P^2 - 4Q \neq 0$, and $V_n = \alpha^n + \beta^n$. If n is composite and

$$U_{n-(D/n)} \equiv 0 \pmod{n}, \tag{15}$$

then we call n a Lucas pseudoprime with parameters P and Q [or lpsp (P, Q)].

An odd composite number n is a Euler Lucas pseudoprime with parameters P, Q [elpsp (P, Q)] if $(n, QD) = 1$ and

$$U_{(n-(D/n))/2} \equiv 0 \pmod{n} \text{ if } (Q/n) = 1, \tag{16}$$
or
$$V_{(n-(D/n))/2} \equiv 0 \pmod{n} \text{ if } (Q/n) = -1. \tag{17}$$

An odd composite number n is a strong Lucas pseudoprime with parameters P, Q [or slpsp (P, Q)] (see [1]) if $(n, D) = 1$ and, with $n - (D/n) = d \cdot 2^s$, d odd, we have either

$$U_d \equiv 0 \pmod{n}, \tag{18}$$
or
$$V_{d \cdot 2^r} \equiv 0 \pmod{n}, \tag{19}$$

for some r with $0 \leqslant r \leqslant s$. Every prime n satisfies the above definitions (with the word "composite" omitted), provided $(n, 2QD) = 1$.

By the theorem of Baillie and Wagstaff [1], if n is a slpsp (P, Q), then n is a elpsp (P, Q).

From Theorem 2 of [19], it follows that if $D = P^2 - 4Q > 0$, $P > 0$, then every arithmetical progression $ax + b$ ($x = 0, 1, 2, \ldots$), where a and b are relatively prime integers, contains an infinite number of odd strong Lucas pseudoprimes with parameters L, Q (that is to say, slpsp for the bases α and β).

If n is prime, and if $(n, Q) = 1$, then

(A) $U_{n-(D/n)} \equiv 0 \pmod{n}$,

(B) $U_n \equiv (D/n) \pmod{n}$,

(C) $V_n \equiv V_1 \equiv \alpha + \beta \pmod{n}$.

I proved in [17] that, if $Q = \pm 1$, $(P, Q) \neq (1, 1)$, then every arithmetical progression $ax + b$ contains an infinite number of odd composite n satisfying (A)-(C), simultaneously [that is to say, pseudoprimes that are at the same time of type (A), (B), and (C)].

First we note that the relations (A)-(C) are linearly dependent (mod n) (see [7]).

Indeed, if $(D/n) = 1$, then

$$AU_{n-1} + B(U_n - 1) + C(V_n - V_1) \equiv 0 \pmod{n}$$

for $A = 2\alpha\beta = 2Q$, $B = -(\alpha + \beta) = -P$

and $C = 1$,

and, if $(D/n) = -1$, then

$$AU_{n+1} + B(U_n + 1) + C(V_n - V_1) \equiv 0 \pmod{n}$$

for $A = -2$, $B = \alpha + \beta = P$, $C = 1$.

Thus, two of the congruences (A), (B), and (C) imply the other one. Now, we whall prove the following theorem.

Theorem 3: If n is a Euler Lucas pseudoprime with parameters P and $Q = \pm 1$, then   satisfies the congruence (A), (B), and (C) simultaneously.

Proof of Theorem 3: If n is A Euler Lucas pseudoprime with parameters P and $Q = \pm 1$, then $(n, D) = 1$ and

$$U_{(n-(D/n))/2} \equiv 0 \pmod{n} \text{ if } (Q/n) = 1,$$

$$V_{(n-(D/n))/2} \equiv 0 \pmod{n} \text{ if } (Q/n) = -1.$$

Since

$$U_{(n-(D/n))/2} \mid U_{n-(D/n)}, \quad V_{(n-(D/n))/2} \mid U_{n-(D/n)},$$

congruence (A) holds.

It remains to prove that $n \mid U_n - (D/n)$. This is true because

$$U_n - (D/n) = U_{(n-(D/n))/2} V_{(n+(D/n))/2} \tag{18}$$

for $n \equiv 1 \pmod{4}$, $Q = -1$ or $Q = 1$, n odd,

and

$$U_n - (D/n) = U_{(n-(D/n))/2} V_{(n-(D/n))/2} \tag{19}$$

for $Q = -1$, $n \equiv 3 \pmod{4}$.

Since every arithmetical progression $ax + b$, where $(a, b) = 1$, contains infinitely many Euler Lucas pseudoprimes with parameters P and Q ($D = P^2 - 4Q > 0$); thus, from Theorem 3 follows the theorem mentioned at the beginning.

Theorem 4: (See also [17].)   Let $Q = \pm 1$, $(P, Q) \neq (1, 1)$, $D = P^2 - 4Q$.   There exist infinitely many composite $n = pq$, where p and q are different primes such that congruences (A), (B), and (C) hold simultaneously.

Definition: A prime p is called a primitive prime factor of the Lehmer number P_n if $p | P_n$ but $p \nmid DLP_3 \cdots P_{n-1}$.

Let $\bar{n}$ denote the square-free kernel of n, that is, n divided by its greatest square factor.   Then the following theorem holds (see [17], [23]).

Theorem: If $(L, Q) = 1$, $\langle L, Q \rangle \neq \langle 1, 1 \rangle$, $\langle 2, 1 \rangle$, $\langle 3, 1 \rangle$ (i.e., β/α is not a root of unity, $\kappa = \overline{Q \max (D, L)}$ and

$$\eta = \begin{cases} 1 & \text{if } \kappa \equiv 1 \pmod 4 \\ 2 & \text{if } \kappa \equiv 2, 3 \pmod 4, \end{cases}$$

$2 \nmid n/\eta Q$ for $Q > 0$, $D > 0$, and $2 \nmid n/\eta \overline{Q \max(D, L)}$ in other cases, then P_n has at least two primitive prime factors for $n > n_0(\alpha, \beta)$.

Proof of Theorem 4: Let $Q = -1$, then $\kappa = \overline{-D} = -\bar{D}$.   If $2 \nmid P$, then $D = P^2 + 4 \equiv 1 \pmod 4$ and $\bar{D} \equiv 1 \pmod 4$, hence $\kappa = 2$. If $4 \| P$, then $D/4 = (P/2)^2 + 1 \equiv 1 \pmod 4$, hence $\bar{D} \equiv 1 \pmod 4$ and also $\kappa = 2$.

Similarly, if $2 \| P$, then $D = P^2 + 4 = 4(4\ell + 1) + 4 = 4(4\ell + 2)$ and $\kappa = 2$.   Thus, if $Q = -1$, then by the theorem in the Definition above, the number $U_{2\bar{D}m}$, where m is odd $> n_0(\alpha, \beta)$ has two primitive prime factors p and q.

If $\alpha\beta = 1$, then U_m for m odd $> n_0(\alpha, \beta)$ has two primitive prime factors p' and q'.

We have

$$p = 2\bar{D}mx + \left(\frac{D}{p}\right), \quad q = 2\bar{D}my + \left(\frac{D}{q}\right), \quad pq - \left(\frac{D}{pq}\right) = 2\bar{D}mz,$$

where x, y, and z are positive integers.

Let $n = pq$, $n - (D/pq) = d \cdot 2^s$, d odd, $2^r \| \bar{D}m$.   We have $\bar{D}m | d \cdot 2^r$, where $0 \leqslant r < s$.   Thus, $V_{\bar{D}m} \equiv 0 \pmod n$, and since $V_{\bar{D}m} | V_{d \cdot 2^r}$, we have $V_{d \cdot 2^r} \equiv 0 \pmod n$, where $0 \leqslant r < s$ and n is slpsp $(P, -1)$ and also elpsp $(P, -1)$, and by Theorem 3 satisfies all three congruences (A), (B), and (C) simultaneously.   In the case $\alpha\beta = 1$, the proof runs in the same way. This completes the proof of Theorem 4.

Examples: Let U_n denote the nth Fibonacci number. The number $U_{10(2k+1)}$ for $k > 1$ has two primitive prime factors. The prime numbers 101 and 151, 71 and 911, 181 and 541, 331 and 39161 are primitive prime factors of U_{50}, U_{70}, U_{90}, and U_{110}, respectively; thus, the numbers $101 \cdot 151$, $71 \cdot 911$, $181 \cdot 541$, $331 \cdot 39161$ are strong Fibonacci pseudoprimes and satisfy the three congruences (A), (B), and (C) at the same time. Let

$$V_n \equiv V_1 \pmod{n}, \quad U_n \equiv (D/n) \pmod{n}$$

For $(D/n) = 1$, we have $\alpha^n + \beta^n \equiv \alpha + \beta \pmod{n}$, $(\alpha^n - \beta^n)/(\alpha - \beta) \equiv 1 \pmod{n}$; hence, $\alpha^n + \beta^n \equiv \alpha + \beta \pmod{n}$, $\alpha^n - \beta^n \equiv \alpha - \beta \pmod{n}$, and $2\alpha^n \equiv 2\alpha \pmod{n}$, $2\beta^n \equiv 2\beta \pmod{n}$, $\alpha^n \equiv \alpha \pmod{n}$, $\beta^n \equiv \beta \pmod{n}$. For $(D/n) = -1$, we have $(\alpha^n - \beta^n)/(\alpha - \beta) \equiv -1 \pmod{n}$; hence, $\alpha^n - \beta^n \equiv -\alpha + \beta \pmod{n}$, and since $\alpha^n + \beta^n \equiv \alpha + \beta \pmod{n}$, we have $\alpha^n \equiv \beta \pmod{n}$, $\beta^n \equiv \alpha \pmod{n}$.

In particular, the case $\alpha = 3$, $\beta = 2$ leads to the problem of Duparc, whether there exist infinitely many composite n for which at the same time we have $3^n \equiv 3 \pmod{n}$, $2^n \equiv 2 \pmod{n}$.

The congruences $\alpha^n \equiv \alpha$, $\beta^n \equiv \beta \pmod{n}$ with $(D/n) = 1$ or $\alpha^n \equiv \beta$, $\beta^n \equiv \alpha \pmod{n}$ with $(D/n) = -1$ imply that, with $(D/n) = \pm 1$, we have $Q^n \equiv Q \pmod{n}$ (since $\alpha\beta = Q$). But if n and Q are given and if $Q \neq \pm 1$, then $Q^n \equiv Q \pmod{n}$ holds very rarely.

In connection with the above, Baillie and Wagstaff [1] asked if there are infinitely many composite n satisfying (A), (B), and (C) for a given $Q \neq \pm 1$. The first such n with $(D/n) = -1$ is $n = 51$, with $P = \pm 17$, $Q = 35$, and $P = \pm 24$, $Q = 16$ (see [1]). Many problems that are solved for ordinary pseudoprimes still hold true for Fibonacci pseudoprimes and have not yet been solved.

PROBLEMS

1. Do there exist infinitely many twin Fibonacci pseudoprimes? (One pair of ordinary pseudoprimes is $17 \cdot 257$ and $17 \cdot 257 + 2$.)
2. Do there exist infinitely many arithmetical progressions formed of four Fibonacci pseudoprimes? (The answer is positive for ordinary pseudoprimes [18].)
3. Do there exist infinitely many Fibonacci pseudoprimes which are at the same time triangular? (For ordinary pseudoprimes, the answer is positive [14].)

4. Do there exist infinitely many Fibonacci pseudoprimes
    which are at the same time pentagonal? (See [15].)
5. Do there exist infinitely many odd natural numbers n
    such that $U_{n-2} \equiv 0 \pmod{n}$? (There exist infinitely
    many positive integers $n > 2$ that satisfy the congru-
    ence $2^{n-2} \equiv 1 \pmod{n}$, but I do not know of any posi-
    tive integer n such that $2^{n-2} \equiv 1 \pmod{n}$ which is less
    than $2^{n_0} - 1$, where $n_0 = 4700063497 = 19 \cdot 47 \cdot 5263229$;
    see [21].)
6. Do there exist infinitely many geometric progressions
    consisting of three different Fibonacci pseudoprimes?

REFERENCES

[1] Baillie, R., and Wagstaff, S., Jr. 'Lucas Pseudo-
     primes.' *Math. Comp.* 35 (1980):1391-1417.
[2] Ko, Chao. 'On the Diophantine Equation $x^2 = y^n + 1$,
     $xy \neq 0$.' *Scientia Sinica* (Notes) 14 (1965):457-460.
[3] Cipolla, M. 'Sui numeri composti P che verificiano la
     congruenza di Fermat $a^{P-1} \equiv 1 \pmod{P}$.' *Annali di
     Matematica* 9 (1904):139-160.
[4] Duparc, H. J. A. *On Almost Primes of the Second Order*,
     pp. 1-13. Amsterdam: Rapport ZW. 1955-013, Math. Cen-
     ter, 1955.
[5] Jarden, Dov. *Recurring Sequences: A Collection of
     Papers.* Jerusalem: Riveon Lematematika, 1958.
[6] Lehmer, Emma. 'On the Infinitude of Fibonacci Pseudo-
     primes.' *The Fibonacci Quarterly* 2, no. 3 (1964):229-
     230.
[7] Lieuwens, E. 'Fermat Pseudoprimes.' Doctoral disser-
     tation, Delft, 1971.
[8] Mordell, L. J. *Diophantine Equations*, pp. 302-304.
     New York and London: Academic Press, 1969.
[9] Nagell, T. 'Sur l'impossibilité de l'equation indé-
     terminée $(x^5 - y^5)/(x - y) = 5z^2$.' *Norsk. Matem. Tids-
     skrift* 1 (1919):51-54.
[10] Parberry, E. A. 'On Primes and Pseudo-primes Related
     to the Fibonacci Sequence.' *The Fibonacci Quarterly* 8
     (1970):49-60.
[11] Pomerance, C.; Selfridge, J. L.; and Wagstaff, S. S.
     'The Pseudoprimes to $25 \cdot 10^9$.' *Math. Comput.* 35 (1980):
     1003-1026.
[12] Van der Poorten, A. J., and Rotkiewicz, A. 'On Strong
     Pseudo-primes in Arithmetic Progressions.' *J. Austral.
     Math. Soc.* Ser. A, 29 (1980):316-321.

[13]   Rotkiewicz, A.   'Sur les nombres pseudopremiers de la forme $ax + b$.'   *C. R. Acad. Sci. Paris* **257** (1963): 2601–2604.

[14]   Rotkiewicz, A.   'Sur les nombres pseudopremiers triangulaires.'   *Elem. Math.* **19** (1964):82–83.

[15]   Rotkiewicz, A.   'Sur les nombres pseudopremiers pentagonaux.'   *Bull. Soc. Roy. Sci. Liége* **33** (1964):261–263.

[16]   Rotkiewicz, A.   'Pseudoprimes, Numbers and Their Generalizations.'   *Student Association of the Faculty of Sciences*, University of Novi Sad, 1972; M.R. 48#8373; Zbl. 324. 10007.

[17]   Rotkiewicz, A.   'On Lucas Numbers with Two Intrinsic Divisors.'   *Bull. Acad. Polon. Sci. Sér. Math. Astr. Phys.* **10** (1962):229–232.

[18]   Rotkiewicz, A.   'The Solution of W. Sierpiński Problem.'   *Rendic. Circ. Mat. Palermo* **28** (1979):62–64.

[19]   Rotkiewicz, A.   'On Euler Lehmer Pseudoprimes and Strong Lehmer Pseudoprimes with Parameters L, Q in Arithmetic Progressions.'   *Math. Comp.* **39** (1982):239–247.

[20]   Rotkiewicz, A.   'Applications of Jacobi's Symbol to Lehmer's Numbers.'   *Acta Arith.* **42** (1983):163–187.

[21]   Rotkiewicz, A.   'On the Congruence $2^{n-2} \equiv 1 \pmod{n}$.'   *Math. Comp.* **43**, no. 167 (1984):271–272.

[22]   Terjanian, G.   'Sur l'equation $x^{2p} + y^{2p} = z^{2p}$.'   *C. R. Acad. Sci. Paris* **285** (1977):973–975.

[23]   Schinzel, A.   'On Primitive Factors of Lehmer Numbers I.'   *Acta Arith.* **8** (1963):213–223.

[24]   Wylie, O.   'Solution of the Problem: In the Fibonacci Series $F_1 = 1$, $F_2 = 1$, $F_{n+1} = F_n + F_{n-1}$, the First, Second, and Twelfth Terms Are Squares. Are There Any Others?'   *Amer. Math. Monthly* **71** (1964):220–222.

[25]   Yorinaga, M.   'A Technique of Numerical Production of a Sequence of Pseudo-prime Numbers.'   *Math. J. Okayama U.* **19** (1976):1–4.

[26]   Yorinaga, M.   'On a Congruencial Property of Fibonacci Numbers—Numerical Experiments.'   *Math. J. Okayama U.* **19** (1976):5–10.

[27]   Yorinaga, M.   'On a Congruencial Property of Fibonacci Numbers—Considerations and Remarks.'   *Math. J. Okayama U.* **19** (1976):11–17.

[28]   Rotkiewicz, A.   'On the Equation $x^p + y^p = z^2$.'   *Bull. Acad. Polon. Sci. Ser. Sci. Math.* **30** (1982):211–214.

Lawrence Somer

LINEAR RECURRENCES HAVING ALMOST ALL PRIMES
AS MAXIMAL DIVISORS

1. INTRODUCTION

This paper generalizes results obtained earlier by the
author [7] concerning second-order linear recurrences having
almost all primes as divisors to higher-order recurrences.
As in the earlier paper, "almost all primes" will be taken
interchangeably to mean either all but finitely many primes
or all primes but for a set of Dirichlet density zero in the
set of primes. The concept of a divisor is generalized to
that of a maximal divisor. Maximal divisors were introduced
by Ward [13], who obtained criteria for a prime to be a maximal divisor of a recurrence. An integer n is a divisor of
a recurrence if n divides some term of the recurrence. A
divisor is a maximal divisor of a kth-order linear recurrence if it divides $k - 1$ consecutive terms of the recurrence, but not more than $k - 1$ consecutive terms.

We shall show that, in general, a kth-order linear
recurrence has almost all primes as maximal divisors if and
only if it contains $k - 1$ consecutive zeros when considered
as a doubly infinite sequence, but no more than $k - 1$ successive zeros. A sequence is called a "k^{th}-order linear
recurrence" if it satisfies a kth-order linear recursion
relation but no lower-order linear recursion relation. Let
(w) be a kth-order linear recurrence defined by the recursion relation

$$w_{n+k} = a_1 w_{n+k-1} + a_2 w_{n+k-2} + \cdots + a_k w_n, \qquad (1)$$

where $a_1, a_2, \ldots, a_k$, and the initial terms $w_0, w_1, \ldots,$
w_{k-1} are all integers. The integers $a_1, \ldots, a_k$ are called
the "parameters" of the recurrence. Associated with the
recurrence (1) is its characteristic polynomial

---

*This paper is based partly on results in the author's Ph.D.
dissertation, The University of Illinois at Urbana-Champaign,
1985.*

A. N. Philippou et al. (eds.), Fibonacci Numbers and Their Applications, 257–272.
© *1986 by D. Reidel Publishing Company.*

$$f(x) = x^k - a_1 x^{k-1} - a_2 x^{k-2} - \cdots - a_{k-1} x - a_k. \tag{2}$$

The characteristic roots r_1, r_2, ..., r_k of (w) and the discriminant D of (w) are the roots and discriminant, respectively, of the characteristic polynomial of (w). It is known by a classical result of finite differences (see, for example, Milne-Thompson [4]) that, if $D \neq 0$, then w_n can be expressed in the form

$$w_n = \sum_{i=1}^{k} c_i r_i^n, \tag{3}$$

where c_i is an element of the algebraic number field $Q(r_1, \ldots, r_k)$ for $1 \leqslant i \leqslant k$.

We distinguish a particular recurrence (u) called the "unit sequence" which satisfies the recursion relation (1) and which has initial terms

$$u_0 = u_1 = \cdots = u_{k-2} = 0, \quad u_{k-1} = 1.$$

Then it is known that

$$u_n = \sum_{i=1}^{k} (1/f'(r_i)) r_i^n, \tag{4}$$

where f is the characteristic polynomial of (u). (See Ward [13].)

Our main result will be Theorem 1.

Theorem 1: Let $k \geqslant 2$. Let (w) be a kth-order linear recurrence over the rational integers with parameters $a_1, \ldots,$ a_k, discriminant D, and characteristic roots $r_1, \ldots, r_k$. Suppose $a_k \neq 0$ and $D \neq 0$. Then almost all primes are maximal divisors of $\{w_n\}_{n=0}^{\infty}$ if and only if there exist $k - 1$ consecutive zeros in the doubly infinite sequence $\{w_n\}_{n=-\infty}^{\infty}$, but no more than $k - 1$ consecutive zeros. Equivalently, almost all primes are maximal divisors of (w) if and only if $\{w_n\}_{n=-\infty}^{\infty}$ is a multiple by a nonzero rational number of a translation of the corresponding unit sequence $\{u_n\}_{n=-\infty}^{\infty}$.

A proof of Theorem 1 for the case $k = 2$ was given in [7]. For this case, we will present a somewhat simpler proof. We call a divisor of a recurrence a null divisor if it divides every term of the recurrence from some point on.

We note that a divisor of a second-order recurrence which is
not a null divisor is a maximal divisor.

The proof of Theorem 1 will depend largely on the following theorem of Schinzel [5].

Theorem 2 (Schinzel): Let K be an algebraic number field.
If λ and θ are nonzero elements of K and the congruence

$$\lambda^x \equiv \theta \pmod{P}$$

is solvable in rational integers for almost all prime ideals
P of K, then the corresponding equation

$$\lambda^x = \theta$$

is solvable for a fixed rational integer.

2. PARTIAL RESULTS

Before proving Theorem 1, we will explore how far we can go
toward proving this theorem using more classical results
than Theorem 2. For a given k^{th}-order recurrence (w) satisfying the recursion relation given in (1), we introduce the
persymmetric determinant $D_n^{(k)}(w)$, defined by

$$D_n^{(k)}(w) = \begin{vmatrix} w_n & w_{n+1} & \cdots & w_{n+k-1} \\ w_{n+1} & w_{n+2} & \cdots & w_{n+k} \\ \vdots & \vdots & & \vdots \\ w_{n+k-1} & w_{n+k} & \cdots & w_{n+2k-2} \end{vmatrix}. \tag{5}$$

By Heymann's theorem (see Milne-Thomson [4], chap. 12.12),

$$D_{n+1}^{(k)}(w) = (-1)^{k+1} a_k D_n^{(k)}(w). \tag{6}$$

By a theorem of Kronecker, the recurrence (w) satisfies a
linear recursion relation of order k, but satisfies no
lower-order linear recursion relation if and only if

$$D_0^{(k)}(w) \neq 0.$$

If neither $D_0^{(k)}(w)$ nor the discriminant D of (w) equals 0,
then, by a result of Ward [12],

$$D_0^{(k)}(w) = c_1 c_2 \cdots c_k D, \tag{7}$$

where the c_i's are the coefficients defined in (3) for $1 \leqslant i \leqslant k$.

The following theorem makes use of the Tchebotarev density theorem and gives constraints on recurrences having almost all primes as maximal divisors.

Theorem 3: Let $k \geqslant 2$. Let (w) be a kth-order linear recurrence over the rational integers with parameters $a_1, \ldots, a_k$. Let φ be a primitive k^{th} root of unity,

$$c = \sqrt[k]{(-1)^{k(k-1)/2} D_0^{(k)}(w)},$$

and

$$d = \sqrt[k]{(-1)^{k+1} a_k}.$$

Let K be the Galois extension field $Q(\varphi, c, d)$. Suppose there exists an automorphism $\sigma \in \text{Gal}(K/Q)$ such that

$$\sigma(\varphi) = \varphi, \quad \sigma(c) \neq c, \quad \text{and} \quad \sigma(d) = d.$$

Then there exists a set of rational primes of positive Dirichlet density which does not contain any maximal divisors of (w).

Proof: Let $G = \text{Gal}(K/Q)$. Suppose σ has h conjugates in G. Then, by the Tchebotarev density theorem, the set of primes p that has a prime ideal divisor P in K whose Frobenius automorphism is σ has a Dirichlet density equal to $h/|G|$. Let p be such a rational prime and suppose that p is also a maximal divisor of (w). Since there are only a finite number of primes dividing the discriminant of K over Q, we can assume that p does not ramify in K. Suppose that $w_n \equiv 0$, $w_{n+1} \equiv 0, \ldots, w_{n+k-2} \equiv 0$, $w_{n+k-1} \not\equiv 0 \pmod{p}$ for some natural number n. It follows by (6) that

$$(-1)^{k(k-1)/2} D_n^{(k)}(w)$$

$$= [(-1)^{k+1} a_k]^n (-1)^{k(k-1)/2} D_0^{(k)}(w). \tag{8}$$

Since $w_n \equiv w_{n+1} \equiv \cdots \equiv w_{n+k-2} \equiv 0 \pmod{p}$, it follows from (5) that

$$(-1)^{k(k-1)/2} D_n^{(k)}(w) \equiv w_{n+k-1}^k \pmod{p}.$$

Thus, $(-1)^{k(k-1)/2}D_n^{(k)}(w)$ is a kth power modulo p. It follows from the definition of σ that $(-1)^{k+1}a_k$ is a kth power modulo p. Hence $[(-1)^{k+1}a_k]^n$ is a kth power modulo p. However, by the definition of σ, the polynomial

$$x^k - (-1)^{k(k-1)/2}D_0^{(k)}(w)$$

does not split into linear factors modulo p. Thus

$$(-1)^{k(k-1)/2}D_0^{(k)}(w)$$

is not a kth power modulo p. Hence, the right side of (8) is not a kth power modulo p, since it is the product of a kth power and a non-kth power modulo p. This is a contradiction, since we have shown earlier that the left side of (8) is a kth power modulo p. The theorem now follows.

Unfortunately, there are recurrences that are not multiples of translations of unit sequences and which do not satisfy the hypotheses of Theorem 3. For example, consider the second-order recurrence (w) with parameters 3, 5, and initial terms 5, 21, 88, 369. Then

$$(-1)^{2(1)/2}D_0^{(2)}(w) = w_1^2 - w_0w_2 = 1,$$

and the conditions of Theorem 3 are not met. However, it is easily seen that this recurrence is not a multiple of a translation of the unit sequence with parameters 3, 5.

3.   PRELIMINARY LEMMAS FOR THE MAIN THEOREM

The following six lemmas will be needed for the proof of Theorem 1.

Lemma 1: Let p be a rational prime. Let (w) be a kth-order linear recurrence over the rational integers with parameters $a_1, \ldots, a_k$. If $a_k \not\equiv 0 \pmod{p}$, then (w) is purely periodic modulo p.

Proof: This is proved by Carmichael [1, p. 344]. The proof follows from the observation that, if $(a_k, p) = 1$, then the term w_n is determined uniquely modulo p by the recursion relation for (w) from the k succeeding terms $w_{n+1}, w_{n+2}, \ldots, w_{n+k}$.

Lemma 2: Let K be an algebraic number field and let R be its ring of integers. Let $\varphi \in R$ be a dth root of unity. Let P be a prime ideal in R. If $d > 1$, suppose that $P \nmid 1 - \varphi^i$ for $1 \leqslant i \leqslant d - 1$. Suppose that

$$\varphi^m = \alpha,$$

where m is a rational integer and $\alpha \in K$. Then, if

$$\varphi^n \equiv \alpha \pmod{P}$$

for a rational integer n, then

$$\varphi^n = \alpha.$$

Proof: It is clear that the lemma holds if $d = 1$. Now suppose that $d > 1$. Since there are only a finite number of values that φ^n can take, one sees that the result will follow if it can be shown that

$$\varphi^i \neq \varphi^j \pmod{P}$$

for $0 \leqslant i < j < d$. Suppose that

$$\varphi^i \equiv \varphi^j \pmod{P}.$$

Then

$$P \mid \varphi^i(\varphi^{j-i} - 1).$$

Since φ^i is a unit in R, we must have that

$$P \mid \varphi^{j-i} - 1.$$

This is impossible by our hypotheses. This contradiction establishes our lemma.

Lemma 3: Let K be an algebraic number field. Let $\alpha \in K$. Let p be a rational prime. Then either $x^p - \alpha$ is irreducible over K or α has a pth root in K.

Proof: This is proved by Van der Waerden [10, p. 171].

Lemma 4: Let K be an algebraic number field. Let $\alpha \in K$. Let p be a rational prime such that α has no p^{th} root in K.

Let φ_p be a primitive p^{th} root of unity. Then α has no p^{th} root in $K(\varphi_p)$.

Proof: Since φ_p satisfies the polynomial

$$x^{p-1} + x^{p-2} + \cdots + x + 1,$$

$[K(\varphi_p) : K] \leqslant p - 1$. Suppose α has no p^{th} root in K. Let β be a pth root of α. Then, by Lemma 3,

$$[K(\beta) : K] = p.$$

If $\beta \in K(\varphi_p)$, then $K(\varphi_p)$ is clearly the splitting field for the polynomial $x^p - \alpha$. However, then $K(\beta) \subseteq K(\varphi_p)$, which is impossible since

$$[K(\beta) : K] = p > p - 1 \geqslant [K(\varphi_p) : K].$$

Thus, if α has no p^{th} roots in K, then α has no p^{th} roots in $K(\varphi_p)$.

Remark: Lemma 4 is known and is given in equivalent form as Satz 16 in [9, p. 298]. For another proof of Lemma 3 and a sketch of a proof of Lemma 4 for the case in which $K = Q$, see Gaal [2, p. 235].

Lemma 5: Let K be an algebraic number field and let R be its ring of integers. Let $\alpha_i \in K$, $\theta_i \in K$, and $m_i \in Z$ for $i = 1$, 2. Suppose neither θ_1 nor θ_2 is a root of unity. Suppose

$$\theta_i^{m_i} = \alpha_i$$

for $i = 1, 2$. Suppose that for almost all prime ideals P of R there exists a rational integer $n(P)$, depending on P but independent of i, such that

$$\theta_i^{n(P)} \equiv \alpha_i \ (\text{mod } P)$$

for $i = 1, 2$. Then $m_1 = m_2$.

Proof: Suppose $m_1 \neq m_2$. Since neither θ_1 nor θ_2 is a root of unity, we can find an odd rational prime p such that $m_1 \not\equiv m_2 \ (\text{mod } p)$ and neither θ_1 nor θ_2 is a pth power in K. This follows by the unique factorization of nonunit and nonzero ideals in R into the product of prime ideals and the Dirichlet-Minkowski unit theorem, concerning the unique decomposition of units that are not roots of unity into the product of a root of unity and powers of fundamental units. By Lemma 4, neither θ_1 nor θ_2 is a pth power in $K(\varphi_p)$, where φ_p is a primitive p^{th} root of unity.

Since every prime ideal of $K(\varphi_p)$ lies over some ideal of K, it follows by hypothesis that, for almost all prime ideals P' of $K(\varphi_p)$,

$$\theta_i^{n(P)} \equiv \alpha_i \pmod{P'}$$

for $i = 1$, 2, where P' lies over the prime ideal P of K. By the Tchebotarev density theorem and Lemmas 3 and 4, the Dirichlet density of prime ideals P' of $K(\varphi_p)$ which split completely in $K(\varphi_p, \theta_i)/K(\varphi_p)$ is $1/p$ for $i = 1$, 2. Thus, the density of prime ideals splitting completely in $K(\varphi_p, \theta_1)/K(\varphi_p)$ or $K(\varphi_p, \theta_2)/K(\varphi_p)$ is less than or equal to $1/p + 1/p = 2/p$. However, $2/p < 1$, since $p > 2$. Therefore, there exists a set of prime ideals $P' \in K(\varphi_p)$ of positive Dirichlet density equal to at least $1 - 2/p$ such that P' does not split completely or ramify in either

$$K(\varphi_p, \sqrt[p]{\theta_1})/K(\varphi_p) \quad \text{or} \quad K(\varphi_p, \sqrt[p]{\theta_2})/K(\varphi_p),$$

$P' \nmid 1 - \varphi_p$, and

$$\theta_i^{n(P)} \equiv \alpha_i \pmod{P'}$$

for $i = 1$, 2. For a prime ideal P' satisfying the above conditions, φ_p is a primitive p^{th} root of unity modulo P', since $\varphi_p \not\equiv 1 \pmod{P'}$ and $\varphi_p^p = 1 \pmod{P'}$. In this case, we must thus have that $N(P') \equiv 1$ modulo p.

By Kummer's theorem, it follows for all but finitely many prime ideals $P' \in K(\varphi_p)$ that, if P' does not split completely in $K(\varphi_p, \sqrt[p]{\theta_i})/K(\varphi_p)$, then θ_i is not a pth power modulo P' for $i = 1$, 2. If θ_i is not a pth power modulo P' and $N(P') \equiv 1 \pmod{p}$, we then have that

$$\theta^{(N(P') - 1)/p} \not\equiv 1 \pmod{P'},$$

and p divides the order of θ_i modulo P' for $i = 1$, 2. Thus, for a set of prime ideals $P' \in K(\varphi_p)$ of positive density, we must have that p divides the order of $\theta_i \pmod{P'}$ and

$$\theta_i^{m_i} = \alpha_i \equiv \theta_i^{n(P)} \pmod{P'}, \text{ for } i = 1, 2.$$

For a prime ideal P' satisfying the above conditions, let $d_i(P')$ be the order of θ_i modulo P' for $i = 1, 2$. Thus,

$$m_i \equiv n(P) \quad [\text{mod } d_i(P')]$$

for $i = 1, 2$. However, $p \mid d_i(P')$ for $i = 1, 2$. Therefore,

$$m_1 \equiv n(P) \equiv m_2 \pmod{p}.$$

This is a contradiction. Hence, $m_1 = m_2$.

Lemma 6: Let K be an algebraic number field and R be its ring of integers. Let $\theta_i \in K$, $\theta_i \neq 0$, $\alpha_i = \theta_i^{m_i}$, and $m_i \in Z$, for $i = 1, 2$. Suppose that θ_1 is a primitive d^{th} root of unity, but θ_2 is not a root of unity. Suppose further that for almost all prime ideals P of R there exists a rational integer $n(P)$ depending on P but independent of i such that

$$\theta_i^{n(P)} \equiv \alpha_i \pmod{P}$$

for $i = 1, 2$. Then $m_1 \equiv m_2 \pmod{d}$.

Proof: This is a consequence of Theorem 4 in a paper by Schinzel [6, p. 246].

4. PROOF OF THE MAIN THEOREM

We are now ready for the proof of Theorem 1.

Proof of Theorem 1: Since $a_k \neq 0$, one can solve uniquely for w_n in terms of $w_{n+1}, w_{n+2}, \ldots, w_{n+k}$ by means of the recursion relation defining (w). Thus, it makes sense to extend $\{w_n\}_{n=0}^{\infty}$ to negative subscripts. The sufficiency of the theorem now follows from Lemma 1.

To prove necessity, it is clear that it suffices to prove that if (w) has almost all primes as maximal divisors, then $\{w_n\}_{n=-\infty}^{\infty}$ is a rational multiple of a translation of the corresponding unit sequence $\{u_n\}_{n=-\infty}^{\infty}$. By (3), we can express w_n in terms of its characteristic roots $r_1, r_2, \ldots, r_k$:

$$w_n = c_1 r_1^n + c_2 r_2^n + \cdots + c_k r_k^n.$$

Let K be the algebraic extension field $Q(r_1, r_2, \ldots, r_k)$. Let R be the ring of integers of K. By hypothesis,

$$r_1 r_2 \cdots r_k = (-1)^{k+1} a_k \neq 0.$$

Moreover, by (7), $c_1 c_2 \cdots c_k \neq 0$ since (w) is a kth-order linear recurrence and $D \neq 0$.

Since every prime ideal P of K lies over a rational prime p, it follows by hypothesis that for almost all prime ideals P in R, there exists a nonnegative rational integer n such that the following $k - 1$ congruences are simultaneously satisfied modulo P:

$$
\begin{aligned}
c_1 r_1^n \quad + c_2 r_2^n \quad + \cdots + c_k r_k^n \quad &\equiv 0 \ (\text{mod } P) \\
c_1 r_1^{n+1} + c_2 r_2^{n+1} + \cdots + c_k r_k^{n+1} &\equiv 0 \ (\text{mod } P) \\
\vdots \qquad \vdots \qquad\qquad \vdots \qquad\qquad &\qquad \vdots \\
c_1 r_1^{n+k-2} + c_2 r_2^{n+k-2} + \cdots + c_k r_k^{n+k-2} &\equiv 0 \ (\text{mod } P).
\end{aligned}
\tag{9}
$$

Without loss of generality, we can assume $P \nmid D$ since $D \neq 0$ and only finitely many prime ideals P divide D. We can rewrite the congruences in (9) as

$$
\begin{aligned}
c_2 r_2^n \quad + \cdots + c_k r_k^n \quad &\equiv -c_1 r_1^n \quad (\text{mod } P) \\
c_2 r_2^{n+1} + \cdots + c_k r_k^{n+1} &\equiv -c_1 r_1^{n+1} \quad (\text{mod } P) \\
\vdots \qquad\qquad \vdots \qquad\qquad &\qquad \vdots \\
c_2 r_2^{n+k-2} + \cdots + c_k r_k^{n+k-2} &\equiv -c_1 r_1^{n+k-2} (\text{mod } P).
\end{aligned}
$$

Solving for c_2 by Cramer's rule and transforming the Vandermonde determinants that are obtained, we get

$$c_2 \equiv -c_1 (r_1/r_2)^n \ (\text{mod } P)$$

if $k = 2$, or

$$
c_j \equiv \frac{-c_1 \left(\prod\limits_{\substack{i \neq j \\ 1 \leq i \leq k}} r_i^n \right) \prod\limits_{\substack{i,\, s \neq j \\ 1 \leq i < s \leq k}} (r_s - r_i)}{\prod\limits_{2 \leq i \leq k} r_i^n \left(\prod\limits_{2 \leq i < s \leq k} (r_s - r_i) \right)}
$$

$$
\equiv \frac{-c_1 r_1^n \prod\limits_{i \neq 1,\, j} (r_i - r_1)}{r_j^n \prod\limits_{i \neq 1,\, j} (r_i - r_j)} \quad (\text{mod } P)
$$

for $2 \leqslant j \leqslant k$ if $k \geqslant 3$. Thus, for almost all prime ideals P of K, there exists a rational integer n such that

$$(r_1/r_2)^n \equiv -c_2/c_1 \quad (\text{mod } P)$$

if $k = 2$, or

$$(r_1/r_j)^n \equiv -(c_j/c_1) \prod_{i \neq 1, j} (r_i - r_j)/(r_i - r_1) \quad (\text{mod } P)$$

for $2 \leqslant j \leqslant k$ if $k \geqslant 3$. Since none of the r_i's or c_i's, $1 \leqslant i \leqslant k$, is congruent to 0 modulo P, it follows by Theorem 2 that there exists a rational integer m such that

$$(r_1/r_2)^m = -c_2/c_1$$

if $k = 2$ or there exists a rational integer m_j such that

$$(r_1/r_j)^{m_j} = -(c_j/c_1) \prod_{i \neq 1, j} (r_i - r_j)/(r_i - r_1)$$

for $2 \leqslant j \leqslant k$ if $k \geqslant 3$. Thus,

$$c_2 = -c_1 (r_1/r_2)^m \tag{10}$$

if $k = 2$ or

$$c_j = -c_1 (r_1^{m_j}/r_j^{m_j}) \prod_{i \neq 1, j} (r_i - r_1)/(r_i - r_j) \tag{11}$$

for $2 \leqslant j \leqslant k$ if $k \geqslant 3$.

If $k \geqslant 3$, we claim that $m_1, m_2, \ldots, m_k$ can be chosen so that

$$m_2 = m_3 = \cdots = m_k = m \tag{12}$$

for some fixed rational integer m. We first suppose that r_1/r_j is a root of unity for $2 \leqslant j \leqslant k$. Let d_j be the order of r_1/r_j. Since $D \neq 0$, there are no repeated characteristic roots and $r_1/r_j \neq 1$ for $2 \leqslant j \leqslant k$. Thus, there is only a finite number of prime ideals $P \in R$ such that $P | (1 - (r_1/r_j)^d$ for $1 \leqslant d < d_j$. Thus, there exists a prime ideal $P \in R$ such that the simultaneous congruences in (9) are all satisfied modulo P, and $P \nmid 1 - (r_1/r_j)^d$ for $2 \leqslant j \leqslant k$ and $1 \leqslant d < d_j$. It then follows from Lemma 2 that (12) holds.

We now assume that r_1/r_j is not a root of unity for some integer j such that $2 \leqslant j \leqslant k$. It now follows from Lemmas 5 and 6 that (12) holds.

Now, if $k = 2$, then using (3), (10), and (4), we see that

$$w_n = c_1 r_1^n + c_2 r_2^n = c_1 r_1^n - c_1 r_1^m r_2^{n-m}$$
$$= c_1 r_1^m (r_1^{n-m} - r_2^{n-m}) = -c_1 r_1^m \Delta_2 u_{n-m},$$

where $\Delta_2 = r_2 - r_1$. Setting $n = m + 1$, one sees that $c_1 r_1^m \Delta_2$ is an element of Q. Thus, (w) is a rational multiple of a translation of (u).

We now assume that $k \geqslant 3$. It follows from (3), (11), and (12) that

$$w_n = c_1 r_1^n + c_2 r_2^n + \cdots + c_k r_k^n \tag{13}$$
$$= c_1 r_1^n - c_1 \sum_{j=2}^{k} \left[r_j^n (r_1^m / r_j^m) \prod_{i \neq 1, j} (r_i - r_1)/(r_i - r_j) \right].$$

Let

$$t = c_1 r_1^m / \prod_{2 \leqslant i < s \leqslant k} (r_s - r_i).$$

Then, by (4) and (13),

$$w_n = t \left[\left(r_1^{n-m} \prod_{2 \leqslant i < s \leqslant k} (r_s - r_i) \right) \right.$$
$$\left. - \sum_{j=2}^{k} \left((-1)^j r_j^{n-m} \prod_{\substack{1 \leqslant c < d \leqslant k \\ c \neq j, \, d \neq j}} (r_d - r_c) \right) \right]$$
$$= ((-1)^{k+1} t) \sum_{j=1}^{k} (\Delta_k / f'(r_j)) r_j^{n-m} = ((-1)^{k+1} t) \Delta_k u_{n-m},$$

where Δ_k is the Vandermonde determinant

$$\prod_{1 \leqslant i < j \leqslant k} (r_j - r_i).$$

Setting $n = m + k - 1$, we again see that (w) is a multiple by an element of Q of a translation of (u).

5. FURTHER RESULTS

For more thoroughness, we now treat the case for which $k = 0$ or 1 and the case for which $a_k = 0$. A linear recurrence is defined to be of order 0 if all its terms are 0.

Theorem 4: Let (w) be a linear recurrence of order k, where $k = 0$ or 1. If $k = 0$, then all primes are null divisors of (w). If $k = 1$, then almost all primes are divisors of (w) if and only if $a_1 = 0$. In this case, all primes are null divisors of (w).

Proof: If $k = 0$, then all terms of (w) are equal to 0 and the theorem follows immediately. If $k = 1$, then $w_0 \neq 0$. Otherwise, $w_n = 0$ for all $n \geq 0$, and (w) satisfies a 0th-order linear recursion relation. If $w_0 \neq 0$, then $w_n = a_1^{n-1} w_0$, and w_n has only those primes which divide w_0 or a_1 as divisors. The theorem now follows.

Theorem 5: Let (w) be a linear recurrence of order k, $k \geq 2$, with characteristic polynomial

$$f(x) = x^m f_1(x),$$

where $m \geq 1$, $f_1(0) \neq 0$, and $f_1(x)$ has no repeated roots. If $m = k$, then all primes are null divisors of (w). If $m = k - 1$, then almost all primes are divisors of (w) if and only if at least one of the terms $w_0, w_1, \ldots, w_{k-1}$ is equal to 0. If $w_{k-1} = 0$, then all primes are null divisors of (w). Finally, suppose $m \geq k - 2$. We say that a divisor of a recurrence is of order r if it divides r consecutive terms of the recurrence but no more than r consecutive terms. Then almost all primes are divisors of order $k - m$ if and only if either exactly $k - m$ consecutive terms among $w_0, w_1, \ldots, w_{m-1}$ are equal to 0 or $\{w_n\}_{n=m}^{\infty}$ is a multiple by a nonzero element of Q of a translation of $\{u_n\}_{n=-\infty}^{\infty}$.

Proof: By hypothesis,

$$a_k = a_{k-1} = \cdots = a_{k-m+1} = 0$$

and $a_{k-m} \neq 0$. If $m = k$, then $w_n = 0$ for $n \geq k$. Consequently, all primes are null divisors of (w). If $1 \leq m \leq k - 1$, then $\{w_n\}_{n=m}$ satisfies a $(k-m)^{\text{th}}$-order recursion relation of the form

$$w_{n+k-m} = a_1 w_{n+k-m-1} + a_2 w_{n+k-m-2} + \cdots + a_{k-m} w_n,$$

where, for $1 \leq i \leq k - m$, a_i is the coefficient of x^i in $f(x)$. By hypothesis, $a_{k-m} \neq 0$. However, this case has been treated in Theorems 1 and 4. By the results of these theorems, we are done.

6. GENERAL RESULTS

As a counterpoise to Theorem 1, which states that essentially only one class of kth-order linear recurrences has almost all primes as divisors, we present results showing that, in general, every k^{th}-order recurrence over the rational integers has an infinite number of prime divisors. Call a recurrence degenerate if any of its characteristic roots is zero or if some ratio of distinct characteristic roots is a root of unity. Ward [14] and [15] showed that nondegenerate second- and third-order linear recurrences with distinct characteristic roots over the rational integers have an infinite number of distinct prime divisors. Laxton [3] proved the following theorem extending Ward's results to k^{th}-order recurrences.

Theorem 6: Let (w) be a nondegenerate, k^{th}-order linear recurrence with distinct characteristic roots over the rational integers, where $k \geqslant 2$. Then (w) has infinitely many distinct prime divisors.

Theorem 7, due to Van Leeuwen [11], generalizes Theorem 6 by showing that all the characteristic roots of (w) do not necessarily have to be distinct.

Theorem 7: Let (w) be a nondegenerate linear recurrence over the rational integers of order $k \geqslant 2$. Then (w) has infinitely many distinct prime divisors.

The proofs of Theorems 6 and 7, due to Laxton and Van Leeuwen, cannot immediately be generalized to prime ideal divisors of kth-order linear recurrences over the rings of integers of arbitrary algebraic number fields. The reason is that the proofs depend on there being only a finite number of units among the rational integers. The only algebraic number fields having only a finite number of units among their rings of integers are those of the form $Q(\sqrt{-N})$, where N is a nonnegative rational integer. Hence, the proofs can only be generalized to show the existence of infinitely many prime ideal divisors for nondegenerate recurrences over the ring of integers of algebraic number fields of the form $Q(\sqrt{-N})$, where N is a nonnegative rational integer.

7. CONCLUDING REMARK

We note that the methods used in the proof of Theorem 1 can be employed to obtain an immediate generalization of this theorem to the determination of recurrences over the ring of integers of an arbitrary algebraic number field having almost all prime ideals as maximal divisors.

ACKNOWLEDGMENTS

I wish to thank Professor Harald Niederreiter for suggestions simplifying the proof of Theorem 1 for the case in which $k = 2$. I also wish to express my appreciation to Professor Lawrence Washington for communications regarding the statements and proofs of Lemmas 3-5. Finally, I would like to thank the referee for references and suggestions concerning Lemmas 4 and 6 and improvements of Theorem 1.

REFERENCES

[1] Carmichael, R. D. 'On Sequences of Integers Defined by Recurrence Relations.' *Quart. J. Pure Appl. Math.* **48** (1920):343-372.

[2] Gaal, L. *Classical Galois Theory*. 2nd ed. New York: Chelsea, 1973.

[3] Laxton, R. R. 'On a Problem of M. Ward.' *The Fibonacci Quarterly* 12, no. 1 (1974):41-44.

[4] Milne-Thomson, L. M. *The Calculus of Finite Differences*. New York: Macmillan, 1960.

[5] Schinzel, A. 'On Power Residues and Exponential Congruences.' *Acta Arithmetica* 27 (1975):397-420.

[6] Schinzel, A. 'Abelian Binomials, Power Residues and Exponential Congruences.' *Acta Arithmetica* 32 (1977): 245-274.

[7] Somer, L. 'Which Second-Order Linear Recurrences Have Almost All Primes as Divisors?' *The Fibonacci Quarterly* 17, no. 2 (1979):111-116.

[8] Somer, L. 'The Divisibility and Modular Properties of kth-Order Linear Recurrences over the Ring of Integers of an Algebraic Number Field with Respect to Prime

Ideals.'   Ph.D. Dissertation, The University of Illinois at Urbana-Champaign, 1985.

[9]   Tschebotorow, N. G.   *Grundzüge der Galoisschen Theorie*.
Groningen-Djakarta, 1950.

[10] Van der Waerden, B. L.   *Modern Algebra*.   Vol. I.   New
York: Ungar, 1949.

[11] Van Leeuwen, J.   'Reconsidering a Problem of M. Ward.'
In *A Collection of Manuscripts Related to the Fibonacci
Sequence*.   Edited by V. E. Hoggatt, Jr., and Marjorie
Bicknell-Johnson.   Santa Clara, Calif.:   The Fibonacci
Association, 1980.

[12] Ward, M.   'Linear Divisibility Sequences.'   *Trans.
Amer. Math. Soc.* **41** (1937):276-286.

[13] Ward, M.   'The Maximal Prime Divisors of Linear Recurrences.'   *Can. J. Math.* **6** (1954):455-462.

[14] Ward, M.   'Prime Divisors of Second-Order Recurring
Sequences.'   *Duke Math. J.* **21** (1954):607-614.

[15] Ward, M.   'The Laws of Apparition and Repetition of
Primes in a Cubic Recurrence.'   *Trans. Amer. Math.
Soc.* **79** (1955):72-90.

Robert F. Tichy

ON THE ASYMPTOTIC DISTRIBUTION OF
LINEAR RECURRENCE SEQUENCES

The Fibonacci numbers F_n ($n = 0, 1, 2, \ldots$) are defined by the well-known linear recursion

$$F_{n+2} = F_{n+1} + F_n, \quad F_0 = F_1 = 1. \tag{1}$$

More generally, linear recursions of second order,

$$a_{n+2} + pa_{n+1} + qa_n = 0, \quad a_0 = \alpha, \quad a_1 = \beta, \tag{2}$$

are considered in this paper. In the first part, the coefficients p, q and the initial values α, β are assumed to be real numbers. We want to determine p, q, α, β such that the sequence $(a_n)_{n=0}^{\infty}$ of fractional parts $\{a_n\} = a_n - [a_n]$ ($[t]$ denotes the greatest integer not greater than the real number t) is everywhere dense in the unit interval $E = [0, 1]$. If

$$\lim_{N \to \infty} \frac{1}{N} \sum_{n=0}^{N-1} 1_I (\{a_n\}) = |I| \quad \text{for every } I \subseteq E, \tag{3}$$

then $(\{a_k\})_{n=0}^{\infty}$ is obviously everywhere dense in E (where 1_I denotes the indicator function of the subinterval I of E and $|I|$ the length of I). A sequence $(a_n)_{n=0}^{\infty}$ fulfilling (3) is called uniformly distributed modulo 1 (u.d. mod 1) because the probability that $\{a_n\}_{n=0}^{N-1}$ is contained in an arbitrary interval $I \subseteq E$ is asymptotically equal to the length $|I|$ of the interval I. In his famous paper [21], H. Weyl showed that a sequence $(a_n)_{n=0}^{\infty}$ is u.d. mod 1 if and only if

$$\lim_{N \to \infty} \frac{1}{N} \sum_{n=0}^{N-1} f(\{a_n\}) = \int_0^1 f(x)\,dx \tag{4}$$

for all continuous functions f with period 1. Since all continuous functions with period 1 can be uniformly approximated by trigonometric polynomials, it follows that $(a_n)_{n=0}^{\infty}$ is u.d. mod 1 if and only if

A. N. Philippou et al. (eds.), Fibonacci Numbers and Their Applications, 273–291.
© *1986 by D. Reidel Publishing Company.*

$$\lim_{N \to \infty} \frac{1}{N} \sum_{n=0}^{N-1} e^{2\pi i h a_n} = 0 \tag{5}$$

for all integers $h \neq 0$. All basic facts on u.d. sequences can be found in the classical monographs of L. Kuipers and H. Niederreiter [8] and E. Hlawka [5].

The elements a_n of the recurrence sequence (2) can be determined explicitly by the following formula:

$$\text{If } \frac{p^2}{4} - q \neq 0: \qquad a_n = \frac{\alpha\mu - \beta}{\mu - \lambda} \lambda^n + \frac{\beta - \alpha\lambda}{\mu - \lambda} \mu^n$$

$$\text{with } \lambda = -\frac{p}{2} + \sqrt{\frac{p^2}{4} - q},$$

$$\mu = -\frac{p}{2} - \sqrt{\frac{p^2}{4} - q};$$

$$\tag{6}$$

$$\text{If } \frac{p^2}{4} - q = 0: \qquad a_n = \left(\alpha + \frac{\beta - \alpha\lambda}{\lambda} n\right)\lambda^n$$

$$(p, q) \neq (0, 0) \qquad \text{with } \lambda = -\frac{p}{2}.$$

A well-known technique, due to Koksma [6] (compare also [8] and [3]), yields the following metric result.

Theorem 1: Let $(a_n)_{n=0}^{\infty}$ be given by $a_{n+2} + p a_{n+1} + q a_n = 0$, $a_0 = \alpha$, $a_1 = \beta$, $[(p^2/4) - q] > 0$, and λ, μ be defined as in (6) with $\max(|\lambda|, |\mu|) > 1$. Then the sequence $(a_n)_{n=0}^{\infty}$ is uniformly distributed modulo 1 for almost all pairs (α, β) of initial values (in the sense of the Lebesgue measure in $\mathbb{R}^2$).

Remark: The proof of Theorem 1 is very technical and is omitted here. Obviously, Theorem 1 can be generalized to linear recurrences of order $s > 2$. Furthermore, quantitative versions can be proved by the method of Erdös-Koksma [2].

The above metric theorem shows the existence of uniformly distributed linear recurrence sequences, but the argument is not constructive. In the following, we apply a simple method of Salem [16] to construct sequences $(a_n)_{n=0}^{\infty}$ of type (2) that are everywhere dense in E. For this purpose, we take a symmetric polynomial of degree 6:

$$P(x) = x^6 + ax^5 + bx^4 + cx^3 + bx^2 + ax + 1 \qquad (7)$$

with integer coefficients a, b, c such that $P(x)$ is irreducible (over the rationals). The equation $P(x) = 0$ is equivalent to

$$\left(x^3 + \frac{1}{x^3}\right) + a\left(x^2 + \frac{1}{x^2}\right) + b\left(x + \frac{1}{x}\right) + c = 0. \qquad (8)$$

Using the substitution $u = x + (1/x)$, it follows that (8) is equivalent to

$$R(u) := u^3 + au^2 + (b - 3)u + (c - 2a) = 0. \qquad (9)$$

We suppose that $R(u)$ has exactly two real roots σ_1 and σ_2 with $2 < |\sigma_1| < |\sigma_2|$ and one real root σ_3 with $|\sigma_3| < 2$. Then $P(x)$ has exactly four real roots λ, $1/\lambda$, μ, and $1/\mu$ with $1 < |\lambda| < |\mu|$ and two complex roots of modulus 1: τ and $1/\tau = \bar{\tau}$. Such a pair of algebraic numbers (λ, μ) has the following fundamental property.

Proposition: If λ, μ with $1 < |\lambda| < |\mu|$ are two real roots of equation (7) as described above, then the argument $\omega = \arg \tau$ is an irrational number.

Proof: We suppose, on the contrary, that $A + B\omega = 0$ with integers A, B, $(A, B) \neq (0, 0)$. Then

$$e^{2\pi i(A + B\omega)} = 1 \quad \text{or} \quad \tau^B = 1.$$

Since the equation $P(x) = 0$ is irreducible, it is known (compare [20]) that its Galois group is transitive, i.e., there exists an automorphism of the Galois group sending the root τ into the root λ. Thus, the automorphism applied to $\tau^B = 1$ gives $\lambda^B = 1$, a contradiction because of $|\lambda| > 1$. Hence, ω is irrational.

Theorem 2: Let λ, μ with $1 < |\lambda| < |\mu|$ be two real roots of equation (7) as described above and $p = -(\lambda + \mu)$, $q = \lambda\mu$. Then the linear recurrence sequence $(a_n)_{n=0}^{\infty}$,

$$a_{n+2} + pa_{n+1} + qa_n = 0, \quad a_0 = \alpha, \quad a_1 = -\frac{p}{2}\alpha,$$

with $|\alpha| \geqslant 1$, is everywhere dense in the unit interval E.

Proof: The roots of $P(x) = 0$ are λ, $1/\lambda$, μ, $1/\mu$, τ, $1/\tau$, and

$$S = \lambda^n + \frac{1}{\lambda^n} + \mu^n + \frac{1}{\mu^n} + \tau^n + \frac{1}{\tau^n}$$

is a symmetric function in the roots of $P(x)$. Therefore, S must be an integer by the fundamental theorem on symmetric functions, because it is a polynomial in the elementary symmetric functions, and the coefficients of $P(x)$ are integers. Hence (with $\omega = \arg \tau$),

$$S = \lambda^n + \mu^n + \frac{1}{\lambda^n} + \frac{1}{\mu^n} + 2 \cos 2\pi n\omega \equiv 0 \text{ modulo } 1,$$

and so [compare to formula (6)]

$$a_n = \frac{\alpha}{2} \lambda^n + \frac{\alpha}{2} \mu^n \equiv \alpha \cos 2\pi n\omega - \frac{\alpha}{2\lambda^n} - \frac{\alpha}{2\mu^n} \text{ mod } 1.$$

Since

$$\lim_{n \to \infty} \frac{1}{\lambda^n} = \lim_{n \to \infty} \frac{1}{\mu^n} = 0,$$

it is sufficient to prove that $(\alpha \cos 2\pi n\omega)_{n=0}^{\infty}$ is everywhere dense in E. By the above proposition, ω is irrational and so $(n\omega)_{n=0}^{\infty}$ is dense in E. So, for an arbitrary given number $b \in E$, we can choose a subsequence $(n_j\omega)_{j=0}^{\infty}$ that is convergent to the number $c \in E$ with $\cos 2\pi c = b/\alpha$. Hence, the sequence $(\alpha \cos 2\pi n_j\omega)_{j=0}^{\infty}$ is convergent to an arbitrary number $b \in E$, and so $(\alpha \cos 2\pi n\omega)_{n=0}^{\infty}$ is everywhere dense in E.

Remark: More generally we can consider instead of the polynomial $P(x)$ in (7) an arbitrary symmetric polynomial of degree $2k$ with integer coefficients that is irreducible (over the rationals). Then the corresponding polynomial $R(u)$ (obtained by the substitution $u = [x + (1/x)]$) must have exactly k real roots σ_1, σ_2, σ_3, ..., σ_k with $2 < |\sigma_1| < |\sigma_2|$ and $-2 < \sigma_3 < \cdots < \sigma_k < 2$; therefore, the given symmetric polynomial has exactly four real roots, λ, $1/\lambda$, μ, and $1/\mu$, with $1 < |\lambda| < |\mu|$, and $2(k - 2)$ complex roots of modulo 1:

$$\tau_1, \frac{1}{\tau_1} = \overline{\tau_1}, \ldots, \tau_{k-2}, \frac{1}{\tau_{k-2}} = \overline{\tau_{k-2}}.$$

If the arguments $\omega_1 = \arg \tau_1$, ..., $\omega_{k-2} = \arg \tau_{k-2}$ and 1 are linearly independent (over the rationals), then a result similar to Theorem 2 is obtained.

Remark: We give a simple example. Let $P(x)$ be given as in (7) with coefficients a, b, and c such that

$$R(-2) = 2a - 2b + c - 2 > 0,$$

$$R(2) = 2a + 2b + c + 2 < 0.$$

If there exists a prime number p with $a \equiv 0(p)$, $b \equiv 0(p)$, and $c \equiv 0(p)$, then $P(x)$ is irreducible by Eisenstein's criterion (compare to [20]). Hence, in this case, we obtain a dense sequence (a_n) as described in Theorem 2.

The proof of Theorem 2 is based on the fact that the sequence $(\alpha \cos 2\pi\omega n)_{n=0}^{\infty}$ is everywhere dense in E for $|\alpha| \geq 1$. In the following, we shall prove that it is not u.d. mod 1 and so the sequence $(a_n)_{n=0}^{\infty}$ is not u.d. mod 1. For short, we only consider the case $\alpha = 1$. The uniform distribution of $f(\omega n) := \cos 2\pi\omega n$ is equivalent to

$$\lim_{N \to \infty} \frac{1}{N} \sum_{n=0}^{N-1} e^{2\pi i h f(\omega n)} = 0$$

for all integers $h \neq 0$ by Weyl's criterion (5). Because of (4),

$$\lim_{N \to \infty} \frac{1}{N} \sum_{n=0}^{N-1} e^{2\pi i h f(\omega n)} = \int_0^1 e^{2\pi i h f(x)} dx,$$

and for all $h \neq 0$,

$$\int_0^1 e^{2\pi i h (\cos 2\pi x)} dx = J_0(4\pi n) \neq 0,$$

hence $(\cos 2\pi\omega n)_{n=0}^{\infty}$ is not u.d. mod 1. But we are able to compute

$$g(x) := \lim_{N \to \infty} \frac{1}{N} \sum_{n=0}^{N-1} 1_{[0,x)}\{\cos 2\pi\omega n\}. \tag{10}$$

We have

$$g(x) = \lim_{N \to \infty} \left(\frac{1}{N} \sum_{n=0}^{N-1} 1_{[0,x)}(\cos 2\pi\omega n) + \right.$$

$$\left. + \frac{1}{N} \sum_{n=0}^{N-1} 1_{[0,x)}(1 + \cos 2\pi\omega n) \right)$$

(continued)

$$= \lim_{N \to \infty} \frac{1}{N} \sum_{n=0}^{N-1} 1_{\left(\frac{1}{2\pi} \arccos x, \frac{1}{4}\right]}(\{\omega n\}) +$$

$$+ \lim_{N \to \infty} \frac{1}{N} \sum_{n=0}^{N-1} 1_{\left[\frac{3}{4}, 1 - \frac{1}{2\pi} \arccos x\right)}(\{\omega n\}) +$$

$$+ \lim_{N \to \infty} \frac{1}{N} \sum_{n=0}^{N-1} 1_{\left(\frac{1}{2\pi} \arccos (x-1), 1 - \frac{1}{2\pi} \arccos (x-1)\right)}(\{\omega n\}).$$

Since ω is irrational, $\{\omega n\}$ is u.d., and so

$$g(x) = \frac{1}{4} - \frac{1}{2\pi} \arccos x + 1 - \frac{1}{2\pi} \arccos x - \frac{3}{4} +$$

$$+ 1 - \frac{1}{2\pi} \arccos (x-1) - \frac{1}{2\pi} \arccos (x-1)$$

$$= \frac{3}{2} - \frac{1}{\pi} \arccos x - \frac{1}{\pi} \arccos (x-1), \qquad (11)$$

thus the sequence $(\cos 2\pi \omega n)_{n=0}^{\infty}$ has the asymptotic distribution function $g(x)$. Now we give an estimate for the error term in the above computation. For this purpose, we need the fundamental concept of the discrepancy $D_N(a_n)$ of a sequence $(a_n)_{n=0}^{\infty}$ of real numbers:

$$D_N(a_n) = \sup_{0 \leqslant x \leqslant 1} \left| \frac{1}{N} \sum_{n=0}^{N-1} 1_{[0,x)}(\{a_n\}) - x \right|. \qquad (12)$$

Similarly, we define, for any asymptotic distribution function (a.d.f.) $g(x)$ ($0 \leqslant x \leqslant 1$, $g(0) = 0$, $g(1) = 1$, with g increasing on E):

$$D(a_n; g) = \sup_{0 \leqslant x \leqslant 1} \left| \frac{1}{N} \sum_{n=0}^{N-1} 1_{[0,x)}(\{a_n\}) - g(x) \right|. \qquad (13)$$

A sequence $(a_n)_{n=0}^{\infty}$ is u.d. mod 1 if

$$\lim_{N \to \infty} D_N(a_n) = 0$$

and it is asymptotically distributed modulo 1 (a.d. mod 1) to the continuous a.d.f. $g(x)$ if and only if

$$\lim_{N \to \infty} D_N(a_n; g) = 0.$$

The discrepancies defined in (12) and (13) are strongly connected.

Theorem 3: If $(x_n)_{n=1}^N$ denotes a sequence with N elements contained in E and $g(x)$ is a strictly increasing continuous function with $g(0) = 0$, $g(1) = 1$, then

$$D_N(x_n; g) = D_N(g(x_n)).$$

Proof: $D_N(x_n; g) = \sup_{0 \leqslant x \leqslant 1} \left| \frac{1}{N} \sum_{n=1}^{N} 1_{[0, x)}(x_n) - g(x) \right|$

$$= \sup_{0 \leqslant x \leqslant 1} \left| \frac{1}{N} \sum_{n=1}^{N} 1_{[0, g(x))}(g(x_n)) - g(x) \right|$$

$$= \sup_{0 \leqslant y \leqslant 1} \left| \frac{1}{N} \sum_{n=1}^{N} 1_{[0, y)}(g(x_n)) - y \right|$$

$$= D_N(g(x_n)).$$

Remark; As a consequence of this simple theorem, all general results for $D_N(x_n)$ can be generalized to $D_N(x_n; g)$. So we obtain, by Niederreiter's explicit formula (see [8], p. 91),

$$D_N(x_n; g) = \frac{1}{2N} + \max_{i = 1, \ldots, N} \left| g(x_i) - \frac{2i - 1}{2N} \right|,$$

where $(x_n)_{n=1}^N$ is a sequence of numbers $x_1 \leqslant x_2 \leqslant \cdots \leqslant x_N$ contained in E. Furthermore, it follows from W. Schmidt's lower bound for $D_N(x_n)$ (see [17]) that

$$D_N(x_n; g) \geqslant c \frac{\log N}{N}$$

for infinitely many N (with an absolute constant $c > 0$). It should be remarked that E. Hlawka [4] considered an estimate related to Theorem 3.

It is well known that the discrepancy $D_N(\omega n)$ of the sequence $(\omega n)_{n=0}^\infty$ is strongly connected with the diophantine approximation properties of the irrational number ω (compare to [8]). In a recent paper [18], J. Schoißengeier found an explicit formula for $D_N(\omega n)$; the following result is a weaker version of this explicit formula: Let $\omega = [0; t_1, t_2, \ldots]$ be the continued fraction expansion of ω with partial quotients t_i and convergents p_n/q_n. For the natural number N, there exist uniquely determined integers m and b_i, $0 \leqslant i \leqslant m$ with

$$N = \sum_{i=0}^{m} b_i q_i, \quad b_m > 0, \quad 0 \leq b_i \leq t_{i+1}$$

for $0 \leq i \leq m$ and $b_{i-1} = 0$, if $b_i = t_{i+1}$ for $0 < i \leq m$. Then, for $N = b q_m$ ($1 \leq b \leq t_{n+1}$, b an integer):

$$D_N(\omega n) = \frac{b}{N} - (b-1)|q_m \omega - p_m| - \frac{b}{N}|q_m \omega - p_m|. \tag{16}$$

If N is an arbitrary natural number, the following estimate holds:

$$ND_N(\omega n) = \max\left(\sum_{\substack{j=0 \\ j \equiv 0(2)}}^{m} b_j\left(1 - \frac{b_j}{t_{j+1}}\right), \sum_{\substack{j=0 \\ j \equiv 1(2)}}^{m} b_j\left(1 - \frac{b_j}{t_{j+1}}\right)\right) + 0(m), \tag{17}$$

where the 0-constant is an absolute one.

Applying (17), Theorem 4 follows immediately.

Theorem 4: Let $\omega = [0; t_1, t_2, \ldots]$ be a given irrational number and m, t_i, and b_i be given as above. If g denotes the a.d.f.

$$g(x) = \frac{3}{2} - \frac{1}{\pi}\arccos x - \frac{1}{\pi}\arccos(x-1),$$

then

$$ND_N(g^{-1}(\omega n); g) \leq 8 \max\left(\sum_{\substack{j=0 \\ j \equiv 0(2)}}^{m} b_j\left(1 - \frac{b_j}{t_{j+1}}\right), \sum_{\substack{j=0 \\ j \equiv 1(2)}}^{m} b_j\left(1 - \frac{b_j}{t_{j+1}}\right)\right) + 0(m).$$

Remark: In the case $N = b q_m$, one can take the exact formula (16). For arbitrary N, there also exists an exact formula; however, it is very complicated and will be omitted here. The interested reader is referred to [18].

To estimate the g-discrepancy of a recursive sequence (a_n), we must investigate how much the sequences $1/\lambda^n$ and $1/\mu^n$ contribute to the distribution behavior.

Theorem 5: Let $x_1, \ldots, x_N$ and $y_1, \ldots, y_N$ be two finite sequences in E and $g(x)$ ($0 \leq x \leq 1$) be a continuous differentiable asymptotic distribution function with

$$g(0) = 0, \ g(1) = 1, \text{ and } M = \max_{0 \leqslant x \leqslant 1} g'(x).$$

Suppose $\varepsilon_1, \ldots, \varepsilon_N$ are nonnegative numbers such that $|x_n - y_n| \leqslant \varepsilon_n$ for $1 \leqslant n \leqslant N$. Then, for any $\varepsilon \geqslant 0$, we have

$$|D_N(x_n; \ g) - D_N(y_n; \ g)| \leqslant M\varepsilon + \frac{\overline{N}(\varepsilon)}{N},$$

where $\overline{N}(\varepsilon)$ denotes the number of n, $1 \leqslant n \leqslant N$, such that $\varepsilon_n > \varepsilon$.

Proof: The proof is a modification of [8, p. 132]. We obtain for the intervals $I = [0, \ x)$ and $I_1 = [0, \ x + \varepsilon)$:

$$\sum_{n=0}^{N-1} 1_{I_1}(x_n) = Ng(x + \varepsilon) + \delta_1 ND_N(x_n; \ g)$$

with $|\delta_1| \leqslant 1$. It follows that

$$\sum_{n=0}^{N-1} 1_I(y_n) - Ng(x) = \sum_{n=0}^{N-1} 1_{I_1}(x_n) + \overline{N}(\varepsilon) - Ng(x)$$

$$\leqslant N(g(x + \varepsilon) - g(x)) + \delta_1 ND_N(x_n; \ g) + \overline{N}(\varepsilon),$$

and by the mean value theorem,

$$\sum_{n=0}^{N-1} 1_I(y \ ) - Ng(x) \leqslant MN\varepsilon + ND_N(x_n; \ g) + \overline{N}(\varepsilon).$$

Similarly, we obtain

$$\sum_{n=0}^{N-1} 1_I(y_n) - Ng(x) \geqslant -MN\varepsilon - ND_N(x_n; \ g) - \overline{N}(\varepsilon);$$

hence,

$$D_N(y_n; \ g) \leqslant D_N(x_n; \ g) + M\varepsilon + \frac{\overline{N}(\varepsilon)}{N}.$$

Interchanging the rolls of (x_n) and (y_n), the inequality of Theorem 5 follows immediately.

Now we consider the sequence $(a_n)_{n=0}^{\infty}$ with $\alpha = 1$ as described in Theorem 2. But the a.d.f.

$$g(x) = \frac{3}{2} - \frac{1}{\pi} \arccos x - \frac{1}{\pi} \arccos(x - 1)$$

has an unbounded first derivative $g'(c)$, and so Theorem 5 cannot be applied directly. We need the following lemma.

Lemma: For $g(x) = \dfrac{3}{2} - \dfrac{1}{\pi} \arccos x - \dfrac{1}{\pi} \arccos(x - 1)$ and .
$0 < \varepsilon < \dfrac{1}{4}$, $0 \leqslant x < x + \varepsilon \leqslant 1$, we have

$$\left| g(x + \varepsilon) - g(x) \right| \leqslant \sqrt{\varepsilon}.$$

Proof: First we note that

$$\left| g(x + \varepsilon) - g(x) \right|$$

$$\leqslant \max(\varepsilon \sup_{\varepsilon \leqslant x \leqslant 1-\varepsilon} g'(x), \ 1 - g(1 - \varepsilon), \ g(2\varepsilon)),$$

and so, because of $1 - g(1 - \varepsilon) = g(\varepsilon) < g(2\varepsilon)$ and

$$g'(x) = \frac{1}{\pi} \frac{1}{\sqrt{1 + x^2}} + \frac{1}{\sqrt{2x - x^2}}$$

$$\leqslant \frac{2}{\pi} \frac{1}{\sqrt{\varepsilon}} \quad \text{for } \varepsilon \leqslant x \leqslant 1 - \varepsilon,$$

it follows that

$$\left| g(x + \varepsilon) - g(x) \right| \leqslant \max\left(\frac{2}{\pi} \frac{1}{\sqrt{\varepsilon}}, \ g(2\varepsilon) \right).$$

It remains to give an upper bound for $g(u)$ $(0 < u < 1/2)$. Since

$$g(u) = \frac{1}{\pi} \arcsin(u(u - 1) + \sqrt{1 - u^2}\sqrt{2u - u^2}),$$

it follows, with Jordan's inequality

$$\left(\sin y \geqslant \frac{2}{\pi} y \quad \text{for } 0 \leqslant y \leqslant \frac{\pi}{2} \right),$$

$$\sin(\pi g(u)) = u(u - 1) + \sqrt{1 - u^2}\sqrt{2u - u^2} \leqslant \sqrt{2u}$$

$$= \frac{2}{\pi} \pi \sqrt{\frac{u}{2}} \leqslant \sin \pi \sqrt{\frac{u}{2}}.$$

Hence, we obtain $g(u) \leqslant \sqrt{\dfrac{u}{2}}$, and so $g(2\varepsilon) \leqslant \sqrt{\varepsilon}$.

Combining the Lemma with Theorem 5 yields

Theorem 6: Let $x_1, \ldots, x_N$ and $y_1, \ldots, y_N$ be two finite sequences in E and let

$$g(x) = \frac{3}{2} - \frac{1}{\pi} \arccos x - \frac{1}{\pi} \arccos(x - 1).$$

Suppose $\varepsilon_1, \ldots, \varepsilon_N$ are nonnegative members such that $|x_n - y_n| \leq \varepsilon_n$ for $1 \leq n \leq N$. Then, for any ε with $0 < \varepsilon < 1/4$, we have

$$|D_N(x_n; g) - D_N(y_n; g)| \leq \sqrt{\varepsilon} + \frac{\overline{N}(\varepsilon)}{N},$$

where $\overline{N}(\varepsilon)$ denotes the number of n, $1 \leq n \leq N$, such that $\varepsilon_n > \varepsilon$.

Now we apply Theorem 6 with $x_n = a_{n-1}$ ($(a_n)_{n=0}^{N-1}$ defined as in Theorem 2, $\alpha = 1$), $\varepsilon_n = 2/\lambda^n$ and $\varepsilon = 1/N^2$. An elementary calculation shows

$$\overline{N}(\varepsilon) \leq \frac{\log(2/\varepsilon)}{\log|\lambda|} \leq \frac{3 \log N}{\log|\lambda|} \quad (\text{for } N \geq 2).$$

This yields

Theorem 7: Let (a_n) be the linear reucrrence sequence as defined in Theorem 2 with $\alpha = 1$ and

$$g(x) = \frac{3}{2} - \frac{1}{\pi} \arccos x - \frac{1}{\pi} \arccos(x - 1).$$

Then, for any integer $N \geq 2$,

$$|D_N(a_n; g) - D_N(\cos 2\pi\omega n; g)| \leq \frac{3 \log N}{N \log|\lambda|} + \frac{1}{N}.$$

The sequence (a_n) is the sum of two sequences (λ^n) and (μ^n) and so the question arises: How does the a.d.f. of (a_n) depend on the a.d.f. of (λ^n) and (μ^n)? For this purpose, we introduce the two-dimensional a.d.f. $g(x, y)$ of a sequence $(x_n, y_n)_{n=0}^{\infty}$ of points in E^2:

$$\lim_{N \to \infty} \frac{1}{N} \sum_{n=0}^{N-1} 1_{[0, x)}(x_n) 1_{[0, y)}(y_n) = g(x, y)$$

$$(0 \leq x, y \leq 1). \qquad (18)$$

The sequence (x_n, y_n) in E is called a.d. mod 1 to the function g, if g fulfils (18). Similarly to the one-dimensional case, we can introduce the g-discrepancy

$$D_N(x_n, y_n; g) = \sup_{0 \leqslant x, y \leqslant 1} \left| \frac{1}{N} \sum_{n=0}^{N-1} 1_{[0, x)}(x_n) \times \right.$$
$$\left. \times 1_{[0, y)}(y_n) - g(x, y) \right|. \tag{19}$$

Obviously, this concept can be generalized to an arbitrary number of dimensions.

Remark: As a generalization of Theorem 3 we obtain, for $g(x, y) = g_1(x)g_2(y)$ [$g_j(0) = 0$, $g_j(1) = 1$, g_j strictly increasing and continuous on E for $j = 1, 2$] and an arbitrary sequence $(x_n, y_n)_{n=1}^{N}$ of N points in E^2,

$$D_N(x_n, y_n; g) = D_N(g_1(x_n), g_2(x_n)). \tag{20}$$

This yields a lower bound for $D_N(x_n, y_n; g)$ similar to Roth's bound for $D_N(x_n, y_n)$ (compare to [8], p. 105). Now we compute the a.d.f. of the sequence $(\cos 2\pi\omega_1 n; \cos 2\pi\omega_2 n)$:

$$g(x, y) = \lim_{N \to \infty} \frac{1}{N} \sum_{n=0}^{N-1} 1_{[0, x)}(\cos 2\pi\omega_1 n) 1_{[0, y)}(\cos 2\pi\omega_2 n) +$$

$$+ \lim_{N \to \infty} \frac{1}{N} \sum_{n=0}^{N-1} 1_{[0, x)}(\cos 2\pi\omega_1 n) 1_{[0, y)}(1 + \cos 2\pi\omega_2 n) +$$

$$+ \lim_{N \to \infty} \frac{1}{N} \sum_{n=0}^{N-1} 1_{[0, x)}(1 + \cos 2\pi\omega_1 n) 1_{[0, y)}(\cos 2\pi\omega_2 n) +$$

$$+ \lim_{N \to \infty} \frac{1}{N} \sum_{n=0}^{N-1} 1_{[0, x)}(1 + \cos 2\pi\omega_1 n) 1_{[0, y)}(1 + \cos 2\pi\omega_2 n)$$

$$= \lim_{N \to \infty} \frac{1}{N} \sum_{n=0}^{N-1} 1_{(g_1(x), 1/4]}(\{\omega_1 n\}) 1_{(g_1(y), 1/4]}(\{\omega_2 n\}) +$$

$$+ \lim_{N \to \infty} \frac{1}{N} \sum_{n=0}^{N-1} 1_{(g_1(x), 1/4]}(\{\omega_1 n\}) 1_{(g_2(y), 1-g_2(y))}(\{\omega_2 n\}) +$$

$$+ \lim_{N \to \infty} \frac{1}{N} \sum_{n=0}^{N-1} 1_{(g_2(x), 1-g_2(x))}(\{\omega_1 n\}) 1_{(g_1, y), 1/4]}(\{\omega_2 n\}) +$$

$$+ \lim_{N \to \infty} \frac{1}{N} \sum_{n=0}^{N-1} 1_{(g_2(x), 1-g_2(x))}(\{\omega_1 n\}) \times$$
$$\times 1_{(g_1(y), 1-g_1(y))}(\{\omega_2 n\}),$$

with

$$g_1(x) = \frac{1}{2\pi} \arccos x \quad \text{and} \quad g_2(x) = \frac{1}{2\pi} \arccos(x - 1).$$

If 1, ω_1, ω_2 are linearly independent over the rationals, then the sequence $(\omega_1 n, \omega_2 n)_{n=0}^{\infty}$ is u.d. mod 1. Therefore, we have

$$g(x, y) = \left(\frac{1}{4} - g_1(x)\right)\left(\frac{1}{4} - g_1(y)\right) +$$

$$+ \left(\frac{1}{4} - g_1(x)\right)(1 - 2g_2(y)) +$$

$$+ (1 - 2g_2(x))\left(\frac{1}{4} - g_1(y)\right) +$$

$$+ (1 - 2g_2(x))(1 - 2g_2(y))$$

$$= \frac{25}{16} - \frac{5}{4} g_1(x) - \frac{5}{2} g_2(x) - \frac{5}{4} g_1(y) -$$

$$- \frac{5}{2} g_2(y) + g_1(x)g_1(y) + 2g_1(x)g_2(y) +$$

$$+ 2g_2(x)g_1(y) + 4g_2(x)g_2(y). \tag{21}$$

Applying a result of Niederreiter [9], we obtain

Theorem 8: Let $(x_n, y_n) = (\{\cos 2\pi\omega_1 n\}, \{\cos 2\pi\omega_2 n\})$ be a given sequence in E^2, ω_1 and ω_2 are real algebraic numbers with 1, ω_1, ω_2 linearly independent over the rationals. Then, for the a.d.f. $g(x, y)$ defined in (21), the following estimate holds:

$$D_N(x_n, y_n; g) = 0(N^{-1+\varepsilon}) \quad \text{for every } \varepsilon > 0.$$

Remark: Obviously, this result can be extended to an arbitrary number of dimensions.

The following—from probability theory well-known result—allows us to compute the a.d.f. of $(x_n + y_n)$, if the two-dimensional a.d.f. of (x_n, y_n) is known.

Theorem 9: Let $g(x, y)$ be the a.d.f. of (x_n, y_n) such that $\partial^2 g/\partial x^2$ is continuous. Then the a.d.f. $\varphi(z)$ of $(x_n + y_n)$ is given by

$$\varphi(z) = g(z, 1) + \int_z^1 \frac{\partial g}{\partial x}(x, 1 + z - x)dx +$$

$$+ \int_0^1 \frac{\partial g}{\partial x}(x, z - x)dx - \int_0^1 \frac{\partial g}{\partial x}(x, 1 - x)dx.$$

Proof: We have (for $0 \leqslant z \leqslant 1$)

$$\lim_{N \to \infty} \frac{1}{N} \sum_{n=0}^{N-1} 1_{[0,\,z)}(x_n + y_n)$$

$$= \lim_{N \to \infty} \frac{1}{N} \sum_{n=0}^{N-1} 1_{[0,\,z)}(x_n + y_n) +$$

$$+ \lim_{N \to \infty} \frac{1}{N} \sum_{n=0}^{N-1} 1_{[1,\,1+z)}(x_n + y_n).$$

For short, we consider only the first lim; the second can be computed in a similar way. For a given partition $0 = \xi_0$ $\xi_1 < \cdots < \xi_K = 1$ with $|\xi_k - \xi_{k-1}| \leqslant \varepsilon$ ($k = 1, \ldots, K$), we obtain

$$\lim_{N \to \infty} \frac{1}{N} \sum_{n=0}^{N-1} 1_{[0,\,z)}(x_n + y_n)$$

$$\geqslant \lim_{N \to \infty} \frac{1}{N} \left(\sum_{n=0}^{N-1} \sum_{k=1}^{K} 1_{[\xi_{k-1},\,\xi_k)}(x_n) 1_{[0,\,z-\xi_k)}(y_n) \right)$$

$$= \sum_{k=1}^{K} (g(\xi_k,\, z - \xi_k) - g(\xi_{k-1},\, z - \xi_k)),$$

and this expression tends to

$$\int_0^1 \frac{\partial g}{\partial x}(x,\, z - x)\,dx,$$

if ε tends to 0.

Similarly, we obtain

$$\lim_{N \to \infty} \frac{1}{N} \sum_{n=0}^{N-1} 1_{[0,\,z)}(x_n + y_n)$$

$$\leqslant \sum_{k=1}^{K} (g(\xi_k,\, z - \xi_{k-1}) - g(\xi_{k-1},\, z - \xi_{k-1})),$$

and so it follows that

$$\lim_{N \to \infty} \frac{1}{N} \sum_{n=0}^{N-1} 1_{[0,\,z)}(x_n + y_n) = \int_0^1 \frac{\partial g}{\partial x}(x,\, z - x)\,dx.$$

Applying the same arguments to

$$\lim_{N \to \infty} \frac{1}{N} \sum_{n=0}^{N-1} 1_{[1, 1+z)}(x_n + y_n),$$

the theorem follows immediately.

Remark: Combining (21) and Theorem 9, the a.d.f. of the sequence $(\cos 2\pi\omega_1 n + \cos 2\pi\omega_2 n)$ can be determined.

In the last part of this paper, we consider the linear recurrence sequence (2) with integer coefficients p, q and integer initial values α, β. Such a sequence $(a_n)_{n=0}^{\infty}$ of integers is called uniformly distributed mod m (m a positive integer $\geqslant 2$) (u.d. mod m) if and only if

$$\lim_{N \to \infty}\left(\frac{1}{N} \operatorname{card}\{n : 0 \leqslant n \leqslant N - 1, \; a_n \equiv j \pmod{m}\}\right) = \frac{1}{m}$$

(22)

for every $j = 0, \ldots, m - 1$. In the case in which m is a prime number, $(a_n)_{n=0}^{\infty}$ is a sequence with elements a_n in a finite field. Uniformly distributed sequences in finite fields are studied in detail in a series of fundamental papers by Niederreiter et al. [10], [11], and [12]. There, all u.d. linear recurrence sequences of second- and third-order with elements in a finite field are characterized. Furthermore, L. Kiupers and J. S. Shiue [7] and H. Niederreiter [13] studied the sequence $(F_n)_{n=0}^{\infty}$ of Fibonacci numbers (in Neiderreiter's paper the moduli 5^k are considered). A general result in this direction was obtained by Bumby in [1]: he characterized all second-order linear recurrence sequences that are u.d. mod m. As a generalization of Bumby's result, G. Turnwald obtained, in his doctoral dissertation [19], the following theorem concerning sequences (a_n) of algebraic integers. There, a sequence $(a_n)_{n=0}^{\infty}$ with elements a_n contained in the ring of algebraic integers A of the algebraic number field K is called uniformly distributed modulo the ideal M of A (u.d. mod M) if and only if

$$\lim_{N \to \infty}\left(\frac{1}{N} \operatorname{card}\{n : 0 \leqslant n \leqslant N - 1, \; a_n \equiv J \pmod{M}\}\right)$$
$$= \frac{1}{N(M)}$$

(23)

for every residue class J mod M [$N(M)$ denotes the norm of M].

Theorem 10: Let $(a_n)_{n=0}$ be a linear recurrence sequence $a_{n+2} = c_1 a_{n+1} + c_0 a_n$ with elements a_n contained in the ring of algebraic integers of the algebraic number field K. Then it follows that

A. $(a_n)_{n=0}^{\infty}$ is u.d. mod M ($M = \pi P_i^{h_i}$ with pairwise different prime ideals P_i and $h_i \geqslant 1$) if and only if $(a_n)_{n=0}^{\infty}$ is u.d. mod $P_i^{h_i}$ for all i and $N(P_i) \neq N(P_j)$ for $i \neq j$.

B. $(a_n)_{n=0}^{\infty}$ is u.d. mod P^h ($h \geqslant 1$) if and only if:

$N(P) = p$, $c_0 \notin P$, and for

$2 \in P$: $c_1 \in P$, $a_1 \not\equiv a_0$ (mod P), and $c_1 \notin P^2$, $c_0 - 1 \notin P^2$ for $h = 2$, and P not ramified for $h > 2$;

$3 \in P$: $c_1^2 + c_0 \in P$, $2a_1 \not\equiv c_1 a_0$ (mod P), and $c_1^2 + c_0 \notin P^2$ for $h = 2$, and P not ramified for $h > 2$;

$2, 3 \notin P$: $c_1^2 + 4c_0 \in P$, $2a_1 \not\equiv c_1 a_0$ (mod P), and P not ramified for $h \geqslant 2$.

A proof of this result will be given by G. Turnwald in an important paper in *Proc. Amer. Math. Soc.*, 1986. Obviously, Bumby's result [1] is a special case of Theorem 10. In the following, we give a simple proof (due to Turnwald [19]) of a general theorem of Rieger [15] concerning the uniform distribution mod m of sequences $(na_n + b_n)$.

Theorem 11: Let (a_n), (b_n) be two sequences of (rational) integers and $\ell(d)$ denote the length of a period mod d of (a_n) and (b_n). If $(a_n, m) = 1$ for every n and $\ell(d) \not\equiv 0$ (mod d) for all d with $d \equiv 0$ (mod m) ($d > 1$), then $(na_n + b_n)_{n=0}^{\infty}$ is u.d. mod m.

Proof: First we assume that (a_n) and (b_n) are purely periodic mod m. Then also $(na_n + b_n)$ is purely periodic mod m [with period $m\ell(m)$]. By Weyl's criterion for uniform distribution mod m (compare to [8]), we have to show

$$\sum_{n=0}^{d\ell(d)-1} e^{2\pi i(h/d(na_n + b_n))} = 0 \text{ for } d > 1,$$

$$m \equiv 0 \text{ (mod } d\text{)}, (h, d) = 1.$$

But this sum is equal to

$$\sum_{k=0}^{\ell(d)-1} \sum_{j=0}^{d-1} e^{2\pi i(h/d(j\ell(d)+k)a_k+b_k))}$$

$$= \sum_{k=0}^{\ell(d)-1} e^{2\pi i(h/d(ka_k+b_k))} \sum_{j=0}^{d-1} e^{2\pi i(h/d(j\ell(d)a_k))}.$$

Because $(h, d) = (a_k, d) = 1$ and $\ell(d) \not\equiv 0 \pmod{d}$, it follows that $h\ell(d)a_k \not\equiv 0 \pmod{d}$. So the inner sum is equal to 0, and the theorem is proved. If (a_n), (b_n) are purely periodic for $n \geq n_0$, the theorem can be proved as above.

Corollary (Niederreiter [14]): Let α_i, β_j, λ_i, μ_j be integers, $a_n = \Sigma\lambda_i\alpha_i^n$, $b_n = \Sigma\mu_j\beta_j^n$. If $(a_n, m) = 1$ for every n, then $(na_n + b_n)$ is u.d. $\pmod{m}$.

Proof: For every integer b the sequence (b^n) is periodic mod d with period $\varphi(d)$ (φ denotes Euler's φ-function). Since $\varphi(d) < d$ for $d > 1$, we have $\varphi(d) \not\equiv 0 \ [\mathrm{mod}\ \varphi(d)]$, and the Corollary follows from Theorem 11.

Corollary: The recurrence sequence $a_{n+2} = 2\alpha a_{n+1} - \alpha^2 a_n$ is u.d. mod m if and only if $(a_1 - \alpha a_0, m) = (\alpha, m) = 1$.

Proof: Because $a_n = n\alpha^{n-1}(a_1 - \alpha a_0) + a_0\alpha^n$, this Corollary follows from the above.

REFERENCES

[1] Bumby, R. T. 'A Distribution Property for Linear Recurrence of the Second Order.' *Proc. Amer. Math. Soc.* **50** (1975):101–106.

[2] Erdös, P., and Koksma, J. F. 'On the Uniform Distribution Modulo 1 of Lacunary Sequences.' *Indag. Math.* **11** (1949):79–88.

[3] Gerl, P. 'Einige metrische Sätze in der Theorie der Gleichverteilung mod 1.' *J. f. reine u. angew. Math.* **216** (1964):50–66.

[4] Hlawka, E. 'On Some Concepts, Theorems and Problems in the Theory of Uniform Distribution.' *Coll. Math. Soc. J. Bolyai* (1974):97–109.

[5] Hlawka, E. *Theorie der Gleichverteilung.* Mannheim-Wien-Zürich: B.I., 1979.

[6]   Koksma, J. F.   'Ein mengentheoretischer Satz über die
      Gleichverteilung modulo Eins.'   *Composition Math.* **2**
      (1935):250-258.

[7]   Kuipers, L., and Shiue, J. S.   'On the Distribution
      Modulo *m* of Sequences of Generalized Fibonacci Num-
      bers.'   *Tonkang J. Math.* **2** (1971):181-186.

[8]   Kuipers, L., and Niederreiter, H.   *Uniform Distribu-
      tion of Sequences.*   New York: John Wiley & Sons, 1974.

[9]   Niederreiter, H.   'Application of Diophantine Approxi-
      mations to Numerical Integration.'   In *Diophantine
      Approximation and Its Applications*, pp. 129-199.   Ed.
      by C. F. Osgood.   New York: Academic Press, 1973.

[10] Niederreiter, H., and Shiue, J. S.   'Equidistribution
      of Linear Recurring Sequences in Finite Fields.'
      Indag. Math. **80** (1977):317-405.

[11] Niederreiter, H., and Shiue, J. S.   'Equidistribution
      of Linear Recurring Sequences in Finite Fields II.'
      Acta Arith. **38** (1980):197-207.

[12] Niederreiter, H.   'On the Cycle Structure of Linear
      Recurring Sequences.'   *Math. Scand.* **38** (1976):53-77.

[13] Niederreiter, H.   'Distribution of Fibonacci Numbers
      Mod 5 .'   *The Fibonacci Quarterly* **10**, no. 4 (1972):
      373-374.

[14] Niederrieter, H.   'Verteilung von Resten rekursiver
      Folgen.'   *Archiv d. Math.* **34** (1980):526-533.

[15] Rieger, J. G.   'Über Lipschitz-Folgen.'   *Math. Scand.*
      45 (1979):168-176.

[16] Salem, R.   'Algebraic Numbers and Fourier Analysis.'
      Heath Math. Monographs, Boston, Mass., 1963.

[17] Schmidt, W.   'Irregularities of Distribution VII.'
      Acta Arith. **21** (1972):45-50.

[18] Schoiszengeier, J.   'On the Discrepancy of $(n\alpha)$.'
      Acta Arith. (to appear).

[19] Turnwald, G.   'Gleichverteilung in algebraischen Zahl-
      körpern.'   Dissertation, Technische Universität Wien,
      1984.

[20] Van der Waerdan, B. L.   *Algebra I.*   Berlin-Heidelberg-
      New York: Springer-Verlag, 1971.

[21] Weyl, H.   'Über die Gleichverteilung von Zahlen mod.
      Eins.'   *Math. Ann.* **77** (1916):313-352.

Added in Proof: From Theorem 7, it can be derived that

$$D_N(a_{ni}g) \leqslant \frac{3}{\log|\lambda|} \frac{\log N}{N} + \frac{1}{N} + 16 D_N(\omega n),$$

where $D_N(\omega n)$ denotes the usual discrepancy of (ωn); see *Anzeiger d. Österr. Akad. Wiss.* (1985), pp. 33–37.

Tony van Ravenstein, Keith Tognetti,
and Graham Winley

GOLDEN HOPS AROUND A CIRCLE

INTRODUCTION

We consider a point hopping around a circle of unit circumference such that each hop is of constant length α, measured clockwise. The $(j+1)^{th}$ hop is then a total distance of $j\alpha$ from some point 0 taken as the origin. In this way, we generate a total of $N_T = N + 1$ points on the circle, namely, 0, 1, 2, ..., N. Hence, we are concerned with the distribution of the points $\{n\alpha\} = n\alpha \bmod 1$, $n = 0, 1, 2, ..., N$.

Our purpose is to describe the patterns associated with the sequence of points distributed around the circle in the particular case of $\alpha = \tau$. τ is the golden section and is the positive root, $(\sqrt{5} - 1)/2$ of the equation $x^2 + x - 1 = 0$. This sequence is quite rich in patterns. There is the pattern of the gaps, of the gap sizes, of the order of the points, of the age of the points, and so on.

What follows is based on the Steinhaus conjecture which claims that for any irrational value of α there are either two or three gap sizes. This has now been proved and may be referred to as "The Three Gap Theorem." (See, for example, van Ravenstein, Tognetti, and Winley [3], [4], and references [1], [5]–[10].) The particular case where α is equal to τ was originally conjectured by J. Oderfeld. In this case, the two gap sizes occur when N_T is a Fibonacci number; otherwise, there are always three gap sizes.

Here, we attempt to give a heuristic presentation which is more concerned with discovery and description rather than proof.

NOTATION

$F_0 = 0$, $F_1 = 1$, $F_j = F_{j-1} + F_{j-2}$, $j = 2, 3, ...$,

[x] = the largest integer not greater than the real number x.

$\{x\} = x - [x]$, the fractional part of the real number x.

A. N. Philippou et al. (eds.), Fibonacci Numbers and Their Applications, 293–304.
© 1986 by D. Reidel Publishing Company.

$N_T = N + 1$, N_T is the total number of points 0, 1, 2, ..., N.

$d_{j,k}$ = the shortest clockwise distance from point j to point $k = \{(k - j)\alpha\} = \{\{k\alpha\} - \{j\alpha\}\}$.

u_j = the jth point from the origin (0) measured clockwise. In particular, $u_{N+1} = u_0 = 0$.

U_N = the sequence of points on the circle = $(0, u_1, u_2, ..., u_N, 0)$, where $u_k \in \{0, 1, 2, ..., N\}$.

For three consecutive points i, j, k on the circle suc $j = k$ and pre $j = i$:

g_k = the length (measured clockwise) of the gap between the points u_{k-1} and $u_k = d_{\ell,m}$, where $m = u_k$ and $\ell =$ pre $u_k = u_{k-1}$. In particular, $g_{N+1} = d_{i,0}$, where $i = u_N$.

α = age of some gap or the difference between N_T and the total number of points present when the gap was formed. It is also equal to the age of the most recent boundary point.

G_N = the sequence of gap lengths = $(g_1, g_2, ..., g_N, g_{N+1})$.

$S_N(j)$ = the number of gaps of length τ^j belonging to G_N.

$P_i = (q_i, p_i)$, the lattice point.

It is seen that $d_{j,k}$ is the distance from point j to the point k on the circle in a clockwise direction disregarding whole numbers of revolutions and $[(k - j)\alpha]$ is the whole number of revolutions between these points. Since $\{-x\} = 1 - \{x\}$, which also holds for x negative, we have $d_{k,k} = 0$, $d_{j,k} = 1 - d_{k,j}$, and $d_{0,k} = \{k\alpha\}$.
Note that pre $0 = u_N$ and suc $0 = u_1$. Also,

$$\sum_{i=1}^{N+1} g_i = 1 \quad \text{and} \quad \sum_{i=1}^{k} g_i = d_{0,j}$$

when $j = u_k$.
The elements of U_N can also be defined as

$$\{u_k : \{u_{k+1}\alpha\} > \{u_k\alpha\}, 0 \leq k \leq N\}.$$

Thus, with $\alpha = \tau = 0.618...$ and $N = 6$, by enumeration, we have $U_6 = (0, 5, 2, 4, 1, 6, 3, 0)$. Hence, $u_6 = 3$ is the sixth point in a clockwise direction (or the first point in an anticlockwise direction) from the origin and

$$d_{0,3} = \{3\tau\} = 1 - \tau^4 = 1 - d_{3,0}$$
$$= 1 - d_{6,3} = d_{3,6}.$$

Also, suc u_6 = suc 3 = 0 and pre u_6 = pre 3 = 6.
We see that

$$G_6 = (\tau^5, \tau^4, \tau^3, \tau^4, \tau^5, \tau^4, \tau^4)$$

and so the largest gap size is $g_3 = d_{2,4} = \tau^3$. Also,

$$d_{0,3} = 1 - \tau^4 = \sum_{i=1}^{6} g_i \quad \text{and} \quad \sum_{i=1}^{7} g_i = 1.$$

GAP SIZES: A HEURISTIC TREATMENT

By enumeration, the following rules may be established. See van Ravenstein et al. [3], [4] for a more detailed account.

Rule 1: All gap sizes are of the form τ^j.

Rule 2:

(a) When a new point is placed it always divides the oldest of the largest gaps in the ratio $1:\tau$ or $\tau:1$.

(b) The first ratio occurs when the largest Fibonacci number not greater than N has an odd index; otherwise, the second ratio occurs.

Rule 3: If N_T is a Fibonacci number, then:

(a) There are exactly two gap sizes; otherwise, there are three gap sizes.

(b) $S_N(j - 2) = F_{j-1}$, $S_N(j - 1) = F_{j-2}$, where $N_T = F_j$, $j = 2, 3, 4, \ldots$.

Rule 4: $S_N(j - 3) = F_j - N$
$S_N(j - 2) = N - F_{j-2}$
$S_N(j - 1) = N - F_{j-1}$
where $F_{j-1} < N_T \leq F_j$, $j = 3, 4, 5, \ldots$.

Rules 3 and 4 may be proved from rules 1 and 2 using induction. The validity of these rules may be checked by referring to Table 1.

Table 1. $S_N(j)$—The Number of Gaps of Size τ^j against j

N_T $\backslash$ j	0	1	2	3	4	5	6	7	8	9	10
1	1										
2		1	1								
3			2	1							
4			1	2	1						
5				3	2						
6				2	3	1					
7				1	4	2					
8					5	3					
9					4	4	1				
10					3	5	2				
11					2	6	3				
12					1	7	4				
13						8	5				
14						7	6	1			
15						6	7	2			
16						5	8	3			
17						4	9	4			
18						3	10	5			
19						2	11	6			
20						1	12	7			
21							13	8			
22							12	9	1		
23							11	10	2		
24							10	11	3		
25							9	12	4		
26							8	13	5		
27							7	14	6		
28							6	15	7		
29							5	16	8		
30							4	17	9		
31							3	18	10		
32							2	19	11		
33							1	20	12		
34								21	13		
35								20	14	1	

Exponent $\tau^j \rightarrow$

$N + 1 = N_T$ (Total Number of Points)

POINTS ADJACENT TO THE ORIGIN

If we consider the quadrant of the real number plane (x, y), where x and y are nonnegative reals, then α may be represented by the slope of the line $y = \alpha x$. This quadrant also contains the grid (or lattice) made up of the lattice points (q, j), where q and j are nonnegative integers.

If we define

$$\delta(q) = \min(\{q\alpha\}, \ 1 - \{q\alpha\}), \tag{1}$$

then $\delta(q)$ is simply the vertical distance from the point $(q, q\alpha)$ lying on the line $y = \alpha x$ to the nearest lattice point (q, p), where p is the nearest integer to $q\alpha$ defined by $|q\alpha - p| = \min_j |q\alpha - p|$. We thus say that (q, p) is the "closest" point to the line at q. If $p > q\alpha$, then (q, p) is "above" the line; otherwise, it is "below."

From the theory of continued fractions (see Khintchine [2]), it is known that

$$\delta_i = \delta(q_i) = \min_{0 \leqslant q \leqslant N} \delta(q), \tag{2}$$

where p_i/q_i is the ith total convergent of the continued fraction expansion of α with the largest denominator less than or equal to N. p_i/q_i is called a best approximation of the second kind to the number α.

On the number plane, this means that the point $P_i = (q_i, p_i)$ is closer to the line $y = \alpha x$ than any other lattice point (q, p) for all $0 < q \leqslant N$. If P_i lies below the line, then from (2), $\delta_i = \{q_i\alpha\}$. If P_i lies above the line, then $\delta_i = 1 - \{q_i\alpha\}$. Hence,

$$\delta_1 = |\alpha q_i - p_i| = \begin{cases} \{q_i\alpha\}, & p_i < q_i\alpha, \\ 1 - \{q_i\alpha\}, & p_i > q_i\alpha. \end{cases} \tag{3}$$

The Case $\alpha = \tau$

$\alpha = \tau$ has the continued fraction expansion,

$$\{0; \ 1, \ 1, \ 1, \ \ldots\} = \cfrac{1}{1 + \cfrac{1}{1 + \cfrac{1}{1 + \cdots}}}$$

It is easily shown that $p_i/q_i = F_i/F_{i+1}$; $i = 0, 1, 2,$ $\ldots$. Hence, from (3),

$$\delta_i = |F_{i+1}\alpha - F_i|, \tag{4}$$

and using $F_i = (1/\sqrt{5})(1/\tau^i - (-1)^i\tau^i)$ in (4), we see that

$$\delta_i = \tau^{i+1}. \tag{5}$$

Figure 1 is a geometric interpretation of the above relationship. From the elementary theory of continued fractions, it is seen that for lattice points $P_i = (q_i, p_i) = (F_{i+1}, F_i)$ with q_i equal to a Fibonacci number of odd index, pionts $(P_0, P_2, P_4, \ldots)$ lie below the line $y = x\tau$ and hence $\delta_i = \{F_{i+1}\tau\}$. On the other hand, for those points P_i where q_i is equal to a Fibonacci number of even index, points $(P_1, P_3, P_5, \ldots)$ lie above the line $y = x\tau$ and for such points $\delta_i = 1 - \{F_{i+1}\tau\}$.

From (5), we have

$$\{F_{i+1}\tau\} = \begin{cases} \tau^{i+1}, & i \text{ even,} \\ 1 - \tau^{i+1}, & i \text{ odd.} \end{cases} \tag{6}$$

If P_i lies below the line, then from (1), $\delta_i = d_{0,q_i}$. If P_i lies above the line, then $\delta_i = 1 - d_{0,q_i}$. Hence, $q_i = F_{i+1} = $ suc 0 if P_i lies below the line and $q_i = F_{i+1} = $ pre 0 if P_i lies above the line.

It is seen from Figure 1 that for $N = 6$ (or 7), $P_4 = (F_5, F_4)$ is the closest point to the line, and since P_4 lies below the line, $q_4 = F_5 = 5 = $ suc 0. The point $P_3 = (F_4, F_3)$ is the next closest point to our line, and being above it we see that $q_3 = F_4 = 3 = $ pre 0.

Reasoning in this manner, we see that if $N + 1$ points are arranged on the unit circle and $n = \max_j\{j : F_j \leqslant N\}$, then n is the index of the largest Fibonacci number j not greater than N and

$$\left. \begin{array}{l} u_1 = \text{suc } 0 = \begin{cases} F_n, & n \text{ odd,} \\ F_{n-1}, & n \text{ even,} \end{cases} \\[2em] u_N = \text{pre } 0 = \begin{cases} F_{n-1}, & n \text{ odd,} \\ F_n, & n \text{ even.} \end{cases} \end{array} \right\} \tag{7}$$

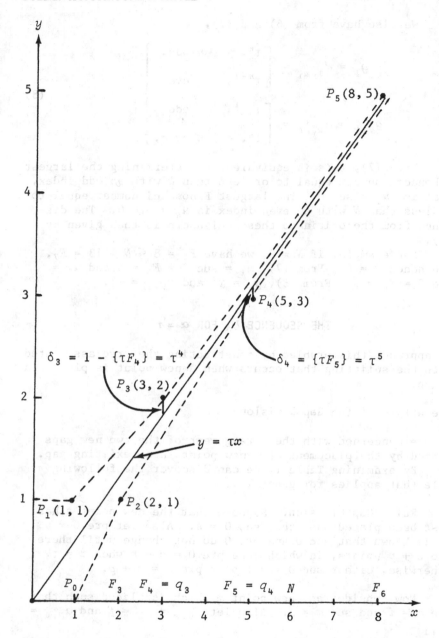

Figure 1.   Geometric representation of δ_i

We also have from (6) and (7),

$$
g_1 = d_{0, u_1} = \begin{bmatrix} \tau^n, & n \text{ odd}, \\[2ex] \tau^{n-1}, & n \text{ even}, \end{bmatrix}
$$

$$
g_{N+1} = d_{u_N, 0} = \begin{bmatrix} \tau^{n-1}, & n \text{ odd} \\[2ex] \tau^n, & n \text{ even}. \end{bmatrix}
$$

$$(8)$$

From (7), this is equivalent to determining the largest Fibonacci number equal to or less than N with an odd index, that is, $u_1 = \text{suc } 0$. The largest Fibonacci number equal to or less than N with an even index is $u_N = \text{pre } 0$. The distance from the origin to these neighbors is then given by (8).

For example, if $N = 9$, we have $F_6 = 8 \leqslant N \leqslant 13 = F_7$, and hence $n = 6$. From (7), $u_1 = \text{suc } 0 = F_5 = 5$ and $u_9 = \text{pre } 0 = F_6 = 8$. From (8), $g_1 = \tau^5$ and $g_{N+1} = \tau^6$.

THE SEQUENCE U_N FOR $\alpha = \tau$

We approach this problem by constructing the tree associated with the splitting that occurs when a new point is placed in a gap.

The Nature of the Gap Division

We are concerned with the arrangement of the two new gaps formed by the placement of a new point in an existing gap.

By examining Table 1, we can discover the following rule that applies for general α.

Rule—Gap Division. Suppose that the new point s has just been placed such that $\text{suc } 0 = s$. Also let $\text{pre } 0 = p$. It is known that $\text{suc } 0$ and $\text{pre } 0$ do not change until there are $s + p$ points, in which case $\text{pre } 0 = s + p$ when $\alpha = \tau$. Otherwise, either $\text{suc } 0 = s + p$ or $\text{pre } 0 = s + p$.

Now consider any new point m which is placed such that $\text{pre } m = i$ and $\text{suc } m = j$. Also let $d_{0, s} = g_1 = f$ and $d_{p, 0} = g_{N+1} = k$.

(a) If m is placed in a gap other than g_1 or g_{N+1}, then $d_{i,m} = f$ and $d_{m,j} = k$.

(b) If m is placed in the gap g_1, then $i = 0$, $j = s$, and $d_{0,m} = f - k$, $d_{m,s} = k$. Henceforth, $d_{0,m}$ becomes the new value of g_1 and m the suc 0.

(c) If m is placed in the gap g_{N+1}, then $i = p$, $j = 0$, and $d_{p,m} = f$, $d_{m,0} = k - f$. Henceforth, $d_{m,0}$ becomes the new value of g_{N+1} and m the new pre 0.

Cases (b) and (c) occur when m is the denominator of a convergent to α. For $\alpha = \tau$, (b) occurs when m is a Fibonacci number of odd index, while (c) occurs when m is a Fibonacci number of even index.

Comment

The gap in which the point s is placed is, of course, the old g_1. It is also the oldest of the largest gaps. The way in which this point is placed is the pattern for the placement of new points in all the other gaps of this size.

Another way of looking at this is to think of the gap division being associated with a switch γ which has a value of either 1 or 2. Thus, if $\gamma = 1$, then a gap of size τ^j fissions into two gaps in clockwise order of τ^{j+1} and τ^{j+2}. If $\gamma = 2$, the order is reversed. Thus, the first clockwise gap after the splitting of τ^j is always $\tau^{j+\gamma}$. This is incorporated into Table 2.

Example: With $N_T = 7$, τ^3 is of age 2 and is the oldest and largest gap. Hence, point 7 will fall in this gap and as $\gamma = 2$ (at $N_T = 7$) the gap will be divided "small:large" (clockwise). At $N_T = 7$, $U_6 = (0, 5, 2, 4, 1, 6, 3, 0)$ and $G_6 = (\tau^5, \tau^4, \tau^3, \tau^4, \tau^5, \tau^4, \tau^4)$ and thus $u_1 = 5$, $u_6 = 3$, and $g_1 = \tau^5$, $g_7 = \tau^4$.

Explanation of Table 2

1. The underlined numbers are the points ordered clockwise. The pair j, a refers to gaps and hence $\underline{V}\ j, a\ \underline{W}$ indicates that the gap of length τ of age a is bounded between points V and W. (V = pre W.)

2. When a new point h is added after $N_T = h$, the following transitions occur:

Table 2. Generator for the Order of Points and Gap Sequence with $\alpha = \tau$

N_T	Y	u_0	g_1	u_1	g_2	u_2	g_3	u_3	g_4	u_4	g_5	u_5	g_6	u_6	g_7	u_7	g_8	u_8	g_9	u_9	g_{10}	u_{10}
1	1	0	0*,0	0																		
2	2	0	1*,0	1	2,0	0																
3	1	0	3,0	2	2,0	1	2,1	0														
4	1	0	3,1	2	2*,1	1	3,0	3	4,0	0												
5	2	0	3*,2	2	3,0	4	4,0	1	3,1	3	4,1	0										
6	2	0	5,0	5	4,0	2	3,1	4	4,1	1	3*,2	3	4,2	0								
7	2	0	5,1	5	4,1	2	3*,2	4	4,2	1	5,0	6	4,0	3	4,3	0						
8	1	0	5,2	5	4,2	2	5,0	7	4,0	4	4,3	1	5,1	6	4,1	3	4*,4	0				
9	1	0	5,3	5	4,3	2	5,1	7	4,1	4	4*,4	1	5,2	6	4,2	3	5,0	8	6,0	0		

(a) $\underline{V}$ j,a $\underline{W}$ to $\underline{V}$ $j,a+1$ $\underline{W}$. Thus, if a gap is not newly formed it simply ages.

(b) $\underline{V}$ j_a^*a $\underline{W}$ to $\underline{V}$ $j+\gamma,0$ $\underline{h}$ $j+\beta,0$ $\underline{W}$. The * indicates that this gap is the oldest of the largest and that it will subdivide such that the first clockwise new gap will be of length $\tau^{j+\gamma}$, where γ is the value of the switch when $N_T = h$. $\beta = 1$ if $\gamma = 2$ and is 2 otherwise. If $n = \max_j\{j : F_j \leqslant N_T\}$, then $\gamma = n \bmod 2 + 1$.

The above method generates points through simple transition rules. An alternative technique developed in [3] results in the following recurrence relationships based on the points adjacent to the origin.

For $n = \max_j\{j : F_j \leqslant N\}$,

$$\text{suc } \hat{j} - j = \left[\begin{array}{ll} F_n, & 0 \leqslant j \leqslant N-F_n, \\ F_{n-2}, & N-F_n < j < F_{n-1}, \\ -F_{n-1}, & F_{n-1} \leqslant j \leqslant N, \\ F_{n-1}, & 0 \leqslant j \leqslant N-F_{n-1}, \\ -F_{n-2}, & N-F_{n-1} < j < F_n, \\ -F_n, & F_n \leqslant j \leqslant N \end{array} \right. \begin{array}{l} \\ \left.\rule{0pt}{3.5em}\right\} n \text{ odd}, \\ \\ \left.\rule{0pt}{3.5em}\right\} n \text{ even}. \end{array} \tag{9}$$

From (9), it may be shown that the following equation holds for $N_T = F_{n+1}$ (and $k = 0, 1, 2, \ldots, F_{n+1}$):

$$u_k = ((-1)^{k+1}kF_n) \bmod F_{n+1}.$$

Using (9) and (6), we can also specify the gap sizes g_k by

$$g_k = \left[\begin{array}{ll} \tau^n, & 0 \leqslant u_{k-1} \leqslant N - F_n, \\ \tau^{n-2}, & N - F_n < u_{k-1} < F_{n-1}, \\ \tau^{n-1}, & F_{n-1} \leqslant u_{k-1} \leqslant N, \\ \tau^{n-1}, & 0 \leqslant u_{k-1} \leqslant N - F_{n-1}, \\ \tau^{n-2}, & N - F_{n-1} < u_{k-1} < F_n, \\ \tau^n, & F_n \leqslant u_{k-1} \leqslant N, \end{array} \right. \begin{array}{l} \\ \left.\rule{0pt}{3.5em}\right\} n \text{ odd}, \\ \\ \left.\rule{0pt}{3.5em}\right\} n \text{ even}. \end{array}$$

REFERENCES

[1]   Halton, J. H.   'The Distribution of the Sequence
      $\{n\xi\}$ $(n = 0, 1, 2, \ldots)$.'   *Proc. Cambridge Philos.
      Soc.* **61** (1965):665–670.

[2]   Khintchine, A. Y. (trans. by P. Wynn). *Continued
      Fractions.* Groningen, The Netherlands: P. Noordhoff,
      1963.

[3]   van Ravenstein, T.; Tognetti, K.; and Winley, G.
      'The Three Gap Theorem.' (In preparation.)

[4]   van Ravenstein, T., Tognetti, K.; and Winley, G.
      'The Golden Section and the Arrangement of Leaves.'
      (Submitted for publication.)

[5]   Révuz, A.   'Sur la Repartition des Points $e^{\nu i \theta}$.'
      C.R. Acad. Sci. **228** (1949):1966–1967.

[6]   Slater, N. B.   'The Distribution of the Integers N
      for Which $\{\theta N\} < \phi$.'   *Proc. Cambridge Philos. Soc.* **46**
      (1950):525–543.

[7]   Slater, N. B.   'Gaps and Steps for the Sequence $n\theta$
      mod 1.'   *Proc. Cambridge Philos. Soc.* **63** (1967):1115–
      1122.

[8]   Sós, V. T.   'On the Theory of Diophantine Approxima-
      tions, I.'   *Acta. Math. Acad. Sci. Hungar.* **8** (1957):
      461–472.

[9]   Sós, V. T.   'On the Distribution mod 1 of the Sequence
      $n\alpha$.'   *Ann. Univ. Sci. Budapest, Eötvös Sect. Math.* **1**
      (1958):127–134.

[10] Świerckowski, S.   'On Successive Settings of an Arc
      on the Circumference of a Circle.'   *Fund. Math.* **46**
      (1958):187–189.